# ROUTLEDGE LIBRARY EDITIONS: 20TH CENTURY SCIENCE

Volume 12

# ATOMIC THEORIES

# ATOMIC THEORIES

F. H. LORING

LONDON AND NEW YORK

First published in 1921
Second edition published in 1923

This edition published in 2014
by Routledge
2 Park Square, Milton Park, Abingdon, Oxfordshire, OX14 4RN

and by Routledge
711 Third Avenue, New York, NY 10017

*Routledge is an imprint of the Taylor and Francis Group, an informa business*

First issued in paperback 2015

*British Library Cataloguing in Publication Data*
A catalogue record for this book is available from the British Library

ISBN 978-0-415-73519-3 (Set)
eISBN 978-1-315-77941-6 (Set)
ISBN 978-1-138-01736-8 (hbk) (Volume 12)
ISBN 978-1-138-96406-8 (pbk) (Volume 12)
ISBN 978-1-315-77928-7 (ebk) (Volume 12)

**Publisher's Note**
The publisher has gone to great lengths to ensure the quality of this book but points out that some imperfections from the original may be apparent.

**Disclaimer**
The publisher has made every effort to trace copyright holders and would welcome correspondence from those they have been unable to trace.

# ATOMIC THEORIES

BY

## F. H. LORING

"No fact discovered about the atom can be
trivial, nor fail to accelerate the progress of
physical science, for the greater part of natural
philosophy is the outcome of the structure and
mechanism of the atom."

Sir J. J. THOMSON
The Romanes Lecture, 1914

WITH 67 FIGURES

SECOND EDITION, REVISED

METHUEN & CO. LTD.
36 ESSEX STREET W.C.
LONDON

*First Published* . . . . . . *October 20th 1921*
*Second Edition, Revised* . . . . *1923*

PRINTED IN GREAT BRITAIN

# PREFACE

THE object of this book is to give the leading facts and theories which relate to the *Atom*, particularly those which have not yet been treated at any length in text-books owing to their newness.

The aim, moreover, in bringing together in one volume the recent work on the atom, is to provide much useful material both for the student and the experimentalist.

In carrying out this aim it seems necessary to include some ideas which have not been thoroughly established. This has the advantage, however, of opening the way for further study and research.

It should be noted that, while the subjects treated involve theory or hypothesis, there are many well-established facts recorded, and the whole trend of the work is to give definition to the ideas hitherto imperfectly understood.

When, for example, using the term *chemical affinity*, it will be seen from these pages that it has a definite meaning attached to it ; and, while the theory involved in this case may not be final, it goes so far in a practical direction as to be extremely useful.

The atomic subjects treated cover a wide range and include the Quantum Theory and Sir J. J. Thomson's recent views of Mass, Energy and Radiation. The Bohr Theory is given at some length, and the Sommerfeld extension of Bohr's work is briefly noted. Prominence is accorded to the recent investigations of Sir E. Rutherford ; to the Octet Theory ; and to Isotopes. The Brownian Movement, including Professor J. Perrin's work, is dealt with in one short chapter, as Perrin's book fully covers the subject. Ionisation Potentials and Solar Phenomena are discussed.

It is not suggested that one particular field of experimental investigation is more important than another, but when certain work has been thoroughly detailed in various text-books—such, for example, as Thomson's Positive-Ray Experiments—the treatment is curtailed here; yet no complete blank is left, as it is desirable to preserve as much continuity of subject as possible without undue overlapping. References are given in order that the reader may follow up any particular branch that elicits special interest.

In describing the Octet Theory, the word *valence* (and *valences*) and its derivatives, as used by Lewis and Langmuir in conformity with American practice, are adopted in order to unify the text where quotations are given. This may seem distasteful to those chemists who have used the word *valency*; but as science is cosmopolitan in character this distinction after all is of small consequence.

It is suggested that the reader may find it desirable not to adhere strictly to the order in which the chapters are arranged in acquainting himself with the various subjects treated.

It may be remarked here that *Atomic Theories* involve the consideration of those phenomena which link up the smallest possible entities with the most potent manifestations of energy known; and on this account such theories should be of interest. It must be remembered too that, however small a given entity may be, the enormous numbers of such entities represent a collective *inertness* or *activity* that is by no means small.

The writer acknowledges the very great assistance given by Councillor J. W. O'Brien in assisting to prepare the MS. for the printers.

F. H. L.

London, *February* 1921

# PREFACE TO THE SECOND EDITION

MOST of the original work of Sir E. Rutherford on the partial breaking up of atoms of low atomic weight has been confirmed and stands, but pending confirmation of some of the experimental results the whole has been left as in the first edition. Among the additions may be mentioned a Periodic Table (frontispiece), a fuller table of isotopes, and the latest atomic weight determinations. A number of corrections have been made.

F. H. L.

LONDON, *October* 1922

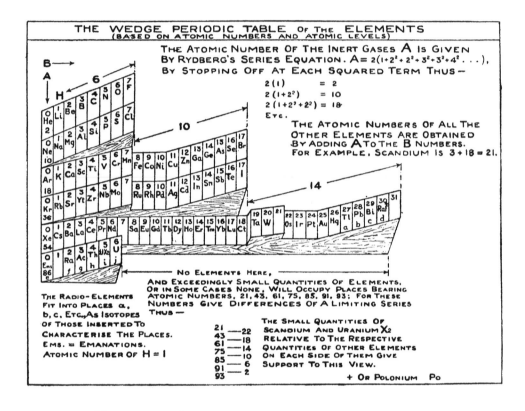

**THE WEDGE PERIODIC TABLE OF THE ELEMENTS**
(BASED ON ATOMIC NUMBERS AND ATOMIC LEVELS)

THE ATOMIC NUMBER OF THE INERT GASES A IS GIVEN BY RYDBERG'S SERIES EQUATION. $A = 2(1 + 2^2 + 2^2 + 3^2 + 3^2 + 4^2 \dots)$, BY STOPPING OFF AT EACH SQUARED TERM THUS —

$$2(1) = 2$$
$$2(1 + 2^2) = 10$$
$$2(1 + 2^2 + 2^2) = 18$$

ETC.

THE ATOMIC NUMBERS OF ALL THE OTHER ELEMENTS ARE OBTAINED BY ADDING A TO THE B NUMBERS. FOR EXAMPLE, SCANDIUM IS $3 + 18 = 21$.

NO ELEMENTS HERE, AND EXCEEDINGLY SMALL QUANTITIES OF ELEMENTS, OR IN SOME CASES NONE, WILL OCCUPY PLACES BEARING ATOMIC NUMBERS, 21, 43, 61, 75, 85, 91, 93; FOR THESE NUMBERS GIVE DIFFERENCES OF A LIMITING SERIES THUS —

THE RADIO-ELEMENTS FIT INTO PLACES α, b, c, ETC., AS ISOTOPES OF THOSE INSERTED TO CHARACTERISE THE PLACES.
EMS. = EMANATIONS.
ATOMIC NUMBER OF H = 1

| | |
|---|---|
| 21 | —22 |
| 43 | —18 |
| 61 | —14 |
| 75 | —10 |
| 85 | —6 |
| 91 | —2 |
| 93 | |

THE SMALL QUANTITIES OF SCANDIUM AND URANIUM $X_2$ RELATIVE TO THE RESPECTIVE QUANTITIES OF OTHER ELEMENTS ON EACH SIDE OF THEM GIVE SUPPORT TO THIS VIEW.

+ OR POLONIUM  Po

NOTES.—Those interested in the problem of classifying the elements may introduce slight changes in some of the atomic levels to suit their ideas of relative similarity between the elements, this diagram being one of first approximation; but such minor changes need not invalidate the usefulness of the scheme. Compare Tables on pages 101 and 102, and see atomic numbers on pages 191 and 193.

# CONTENTS

## CHAPTER I

## CHAPTER II

## CHAPTER III

## CHAPTER IV

## CHAPTER V

## CHAPTER VI

## CHAPTER VII

## CHAPTER VIII

## CHAPTER IX

## CHAPTER X

CHAPTER XXII

CHAPTER XXIII

CHAPTER XXIV

APPENDIX I

APPENDIX II

APPENDIX III

APPENDIX IV

APPENDIX V

APPENDIX VI

APPENDIX VII

APPENDIX VIII

# ATOMIC THEORIES

## CHAPTER I

### INTRODUCTION: ATOMIC THEORIES: EARLIER VIEWS

DEMOCRITUS, at a time dating back to 400 B.C., gave expression to atomic ideas in connection with matter, etc., which in a general sense are true to-day. So striking are the ideas attributed to this philosopher that Tyndall mentioned them, and Millikan * quotes from Tyndall, as follows :—

" 1. From nothing comes nothing. Nothing that exists can be destroyed. All changes are due to the combination and separation of molecules.

" 2. Nothing happens by chance. Every occurrence has its cause from which it follows by necessity.

" 3. The only existing things are the atoms and empty space ; all else is mere opinion.

" 4. The atoms are infinite in number and infinitely various in form ; they strike together and the lateral motions and whirlings, which thus arise, are the beginnings of worlds.

" 5. The varieties of all things depend upon the varieties of their atoms, in number, size and aggregation.

" 6. The soul consists of fine, smooth, round atoms like those of fire. These are the most mobile of all. They interpenetrate the whole body and in their motions the phenomena of life arise."

Professor Millikan remarks : " These principles with a few modifications and omissions might almost pass muster to-day." Of course, the reader will appreciate that it is a far cry from such dreams to the present-day knowledge of matter and electricity.

The first definite step in the development of the atomic theory was made independently by William Higgins and John Dalton—the former in 1789 advanced the idea that each chemical compound was definite in the molecular sense. For instance, the oxides of nitrogen according to Higgins would be : $NO$, $NO_2$, $NO_3$, $NO_4$

---

* *The Electron* (1917), p. 9. The quotation has no bearing upon religious matters. It is not supposed that the ideas quoted were known exactly as stated at the time named, as such information is more or less handed down by various writers from time to time and suffers change in consequence.

and $NO_5$.  Dalton in 1803 arrived at a similar conclusion whereby the nitrogen oxides would be $N_2O$, $NO$ and $NO_2$.*

These results involved the conception of *atoms combining in definite proportions*, but Higgins regarded the $1 : 1$ proportion as the more stable.  It would appear that Dalton had no knowledge of the earlier views of Higgins, which were published in a book,† the title of which would not have invited Dalton's attention.  It is to Dalton that the credit of developing the theory properly belongs ; hence his name has been rightly associated with it.

The atomic theory involves by analysis three principles, viz. (1) *Definite* proportions ;   (2) *Multiple* proportions ;   and (3) *Equivalent* proportions.  In the *first*, the constituents of every given chemical compound are fixed and immutable ; i.e. however formed, and wherever found, a particular compound is the same. It seems safe to remark here that no comprehensive statement can be made which is not liable to slight modification as more knowledge is gained.  For example, the different species of lead (isotopes) differ slightly in atomic weight and, therefore, compounds involving these leads would not contain exactly equal proportionate amounts of lead, though all the compounds would bear the same name and they would be chemically identical.  In the *second*, when one element combines with another, the proportions by weight are in multiples of the smallest proportion corresponding with the atom : in the sense that $A+B=P$ ;  or $2A+B=Q$ ; or $2A+3B=R$, etc., P, Q, R, etc., representing compounds.  In the *third*, each element in combining with or displacing another element follows a rule of fixed proportions by weight represented by a particular number, or a simple multiple or submultiple of that number, for each element involved, which represents the relative combining weight.  The atomic weight then becomes a whole multiple or submultiple of its combining weight.

These regularities led to the establishment of the atomic theory involving definite weights and the conception of valence numbers. In this case the atomic weight divided by the combining weight represents the valence ; or, knowing the valence integer and the combining or equivalent weight, these two quantities multiplied together give the atomic weight.  These matters become clearer if approached with later knowledge, as will be seen from the following chapters.

Dalton regarded the atoms as indivisible particles which for a given element were similar to one another in respect of atomic weight.  While this idea is true in the main, the presence of isotopes introduces a modification in the theory, as will be seen later.

The atom, moreover, cannot be defined as an unalterable entity, even in the chemical sense, for if a hydrogen atom can part with an electron and become thereby a positive ion, according to current

---

* See Meldrum, *Chem. News* (1911), 104, p. 49.
† *A Comparative View of the Phlogistic and Antiphlogistic Theories* (1789).

theory, its character or composition is not, strictly speaking, preserved, as H plus an electron does not equal H minus an electron. Furthermore, recent experiments, as fully detailed in Chapter VIII, show that the atoms of some of the lighter elements can be partly disintegrated by means of a process of bombardment, the dismembered parts having masses of the order of 1, 2 and 3, hydrogen being taken as unity.

The atoms of Dalton have been regarded as the smallest particles of matter which take part in chemical combination, but there are other considerations which make this definition too narrow. The inert elements, such as helium, neon, argon, etc., do not enter into chemical combinations, yet they are made up of atoms. The electron, which is an atom of negative electricity, plays a profound part in the process of chemical combination, and it is also the active agent in electrical phenomena ; and this entity has to be considered in chemical and in physical processes, as will be seen from what follows. The electron in various numbers appears to be an essential constituent of all atoms.

There are other phenomena besides the above which give expres-

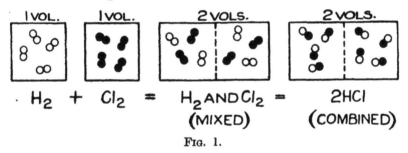

$$H_2 + Cl_2 = H_2 \text{ AND } Cl_2 = 2HCl$$
(MIXED)    (COMBINED)

FIG. 1.

sion to the principle of discontinuity of matter, and some of these lead to the reality of molecules. The kinetic theory of gases affords strong evidence that the molecular state of matter is one which is in harmony with the atomic basis of all compounds, for, in breaking up and manipulating gaseous bodies known to consist of molecules in rapid movement, the gas laws apply and in particular the law of Avogadro stating that the number of molecules in a given volume of any gas is the same when the pressure of the gas on the walls of the vessel is constant and its temperature is also constant. The law of Avogadro can be deduced from the kinetic theory of gases. In applying Avogadro's law when decomposition of gaseous molecules takes place accompanied by recombination to form a different gaseous substance, the atomic theory becomes evident, as will be seen from the well-known example of Fig. 1.

There are five laws, not all of which are exact, that afford an interpretation of gas properties which is in harmony with the atomic theory. These may be briefly stated as follows :—

1. Boyle's law states that the pressure ($p$) multiplied by the

volume ($v$) of a given weight of gas is constant when the temperature is maintained constant : that is, $pv=$ constant.  The following diagram, Fig. 2, after one adopted by Mellor,* should make this law clearer.

As stated above, the gas laws are not exact, and to meet this difficulty van der Waals proposed an equation which gives the behaviour of the gas over a wide range of temperature and pressure. Before coming to van der Waals' equation it will be desirable to proceed a little further.

2. The law of Charles or Gay-Lussac connecting the volume of a gas with its temperature states that under conditions of constant pressure the volume increases by $\frac{1}{273}$rd of its value at 0° C. for every 1° C. increase of temperature.  This law can be stated so as to connect temperature and pressure, for, by keeping the volume constant, for each 1° C. rise of temperature the pressure increases $\frac{1}{273}$rd of that at 0° C.

This is not the place to discuss gas laws, since it belongs more

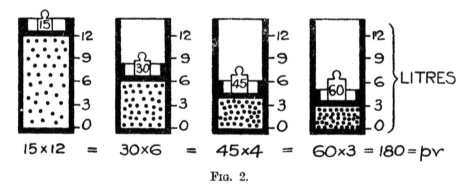

$$15 \times 12 \quad = \quad 30 \times 6 \quad = \quad 45 \times 4 \quad = \quad 60 \times 3 = 180 = pv$$

Fig. 2.

properly to another branch of chemical physics, but van der Waals' reasoning which was the basis of his famous equation is of particular interest in connection with the fundamental behaviour of atoms of which this book treats.

Starting with the general characteristics of all gases, viz. that the molecules are in themselves perfectly elastic and all the energy of a striking blow is returned to the particle as kinetic energy, i.e. energy of motion, the following factors have been taken into account by van der Waals :—

(i) The molecules have definite sizes, and larger molecules will make more to-and-fro excursions between opposite walls of the containing vessel than smaller ones, assuming in both cases their velocities to be the same.  This will be quite obvious if the molecules are imagined to be very large on the one hand and very small on the other, the walls being the same distance apart in both cases. There is in consequence of the increase in molecular size an increase of pressure for, as just shown, more impacts per second take place.

* *Modern Inorganic Chemistry* (1912), p. 78.

This may be expressed in terms of a contracted volume $v-b$, where $b$ is a constant dependent upon the space occupied by the molecules.

(ii) A second correcting factor is based upon the supposition that the molecules attract one another appreciably at high pressures when they are closer together, and van der Waals assumed that the attraction was proportional to the product of the masses of the gas particles, or to the square of the density of the gas ; this factor appears in the form $a/v^2$, $a$ being a constant which varies in passing from one kind of gas to another.

Combining these two factors the final equation becomes

$$\left(p+\frac{a}{v^2}\right)(v-b)=RT$$

in which

> $p=$observed pressure of the gas.
> $v=$measured volume of the gas.
> $a=$constant as above stated.
> $b=$constant as above stated, which is taken as 4 times
>    (or $4\times\sqrt{2}$ times) the actual volume occupied by the molecules.
> $T=$temperature on the absolute scale.
> $R=$gas constant (as in $PV=RT$, which fulfils Avogadro's law, $P$ being the pressure and $V$ the gram-molecule volume of the gas).

By way of a few examples, the above constants for $H_2$, $O_2$ and $N_2$ have the following values :—

> for $a$, 0·00042, 0·00273 and 0·00259 respectively.
> for $b$, 0·00088, 0·00142 and 0·00165 respectively.

Passing on to other gas laws in order to complete this summary—

3. Dalton's law of partial pressures states that the pressure due to a mixture of gases is equal to the sum of the pressures exerted by each gas singly, when confined in the same space as that containing the mixture.

4. Graham's law of diffusion of gases states that the rates of diffusion are inversely proportional to the square roots of their densities.

5. Gay-Lussac's law of combining volumes states that when gases unite the respective volumes initially taken are to each other as small integers (1, 2, 3. etc.), the temperature and the pressure being the same, and the amounts (volumes) being taken so as not to leave any uncombined gas over after the combination has taken place. This law is not necessarily exact, as might be judged from the correction of van der Waals given above.

Returning to Avogadro's law, it will be seen that if in every gas the molecules per cubic centimetre are the same in number at a fixed temperature and pressure * it is only necessary to weigh the vessel containing the gas to compare the molecular weight of its molecules with that of the molecules of another gas in a similar vessel, so long as the temperature and the pressure are kept constant in both cases. By this means it becomes possible to establish the relative weights of molecules of different gases and also of the atoms of the inert gases since they exist in a free state as molecules. It is necessary to point out that corrections have to be made owing to the deviations from ideal gas laws, as indicated above. Having arrived at a standard for comparison it became possible to check chemical methods of determining atomic weights. Oxygen ($O_2$) may be taken, therefore, as a standard having a molecular weight of 32. By means of the chemical and the gas methods the atomic weights of hydrogen and oxygen have been determined independently of each other, with the result that the former by either method is very close to the value 1·008, the probable maximum error not exceeding one unit in the third decimal place. Oxygen has the exact value 16·000, being *exact* of necessity, for it was taken as the standard. As will be seen later, this standard has turned out to be a significant choice, for *all the atomic weights appear to be whole numbers except hydrogen* which has relatively an appreciable fraction. If hydrogen had an approximate atomic weight of 100 it would be necessary to make the exact figure 100·8 for it to have the same *proportionate* fraction as in the case of H=1·008. This fractional value of hydrogen and its position in the Periodic Table raise an interesting point which is discussed in Chapter XX.

The atomicity of electricity itself is another phase of atomic theory, and it is linked up intimately with atomic phenomena as instanced by the valence law established by Faraday, which may be briefly stated as follows :—

In electrolysis it was found that the same quantity of electricity (quantity in the sense of a quart of electricity by analogy, so that this quantity taken as a unit and flowing past a given point in a second represents a current of one ampere) measured in units called *coulombs*, deposits quantities of metal which are proportional to the equivalents, i.e. the atomic weights divided by their valences.

This phenomenon brought to light the fact that a natural unit quantity of electricity is always present in electrolytic actions, and the theory of ions, or carriers of + and − units of electricity from pole to pole in the electrolyte was developed. In other words, the ratio of the coulombs which had passed to the quantity of metal deposited on the cathode must be the same as the ratio of the ionic charge to the mass of the ion. In the case of a simple monovalent ion, this ratio, it will be seen, is inversely proportional to its atomic weight. Hydrogen has the smallest atomic weight, so that the charge associated with hydrogen may be taken as the

---

* The actual number of gas molecules per cubic centimetre at normal temperature (0° C.) and pressure (760 mm.) is $2·705 \times 10^{19}$.

fundamental unit which is now known to be the quantity of electricity represented, or neutralised, by the electron. This last-named fact was not ascertained until the number of hydrogen atoms involved in the electrolytic processes became known. This number was first arrived at approximately by the study of gas laws.

This quantity of electricity varies with different elements, since the same current strength flowing for a given time ($=n$ coulombs) will deposit from silver solution, for example, an amount of silver proportional to its atomic weight; and from a copper solution an amount equal to half the atomic weight of copper, since the cupric ion carries twice as much electricity as the silver ion. Thus it will be seen that there is a natural quantum of electricity and this occurs in whole-number multiples.

The question here might be raised: whether the structure of atoms is always such as to accommodate only a certain minimum quantity of electricity so that its granular nature is due to structural regularities in the atomic edifices which are now known to be complex. There are, however, other experimental facts which point to the atomic nature of electricity, which will appear from a study of the following chapters.

# CHAPTER II

## ATOMIC WEIGHTS AND WHOLE-NUMBER ISOTOPES

PASSING now to more modern thought and work, the long line of experimental researches that followed from the principles laid down by Dalton firmly established the atomic theory of matter. To-day the further problems to be solved are the structure of the atoms themselves and the mechanism which links them together and if possible the utilisation of the energy latent within the atom. There is very often no sharp transition in the nature of the experimental work, as that which has led recently to the isolation and mass measurement of atoms and molecules in the vacuum tube had previously led to the determination of the mass of the much smaller entity : the electron ; this being in a sense a structural unit of the atom. Similarly, the ultra-microscope is an adaptation of the microscope, the former revealing the presence and movement of particles approximately of molecular dimensions (see Chapter XXII).

Professor Sir J. J. Thomson,* by means of a perfected vacuum-tube apparatus, succeeded in photographing the impacts of individual atoms and molecules. Those particles which possessed a charge, usually positive, were deflected by superposed magnetic and electrostatic fields through which they were allowed to stream almost in single file after being passed through a long delivery tube of exceedingly small bore which selected those flying in the right direction. Only the charged particles were deflected by the fields. By measuring the position of the place of their impact—in reality many infinitesimal black spots were recorded on the photographic plate which, being close together, formed a definite line of parabolic curvature for all masses bearing a certain charge to mass ratio— relative to the axes of the system intersecting at the central spot made by the uncharged particles, the masses of the atoms and gas-molecules long known to the chemist were definitely corroborated.† In measuring the photographically-revealed deflections, that is to say, the lines on the plates, a travelling microscope was employed. In a similar way the mass of the electron had been determined. Since the deflected atoms for the most part possessed a positive

---

* *Rays of Positive Electricity* (1913) ; later edition, 1921.

† Calculations involving certain factors are given in Thomson's work here cited. See also J. A. Crowther, *Ions, Electrons and Ionising Radiations* (1919).

charge of electricity the method has been defined as that of *positive rays*. Of course, the electrons themselves being negative entities of electricity are not positive rays, and thereby a further distinction is made by the use of this term.

A recent improvement in the above method by F. W. Aston,* which is more than an improvement in technique as it involves a new method of deflection, has yielded results which have already added considerably to the knowledge of atoms ; for he has been able to record the masses of a number of atoms with an accuracy of about one part in a thousand, and the isotopic character of some of the common elements, hitherto supposed by many to be homogeneous, has been thereby established. The records are obtained photographically and are termed *mass spectra* of atoms and molecules.

Many of the atomic masses, or atomic weights—these terms being used synonymously though there is a fundamental distinction (see Appendix II)—were thought to be shared, as it were, equally by all the atoms of a given element, but they are now being found in many cases to be *mean* values : two (sometimes more, as shown in Table II) chemically identical species † of atoms exist where one species was hitherto believed to exist, being slightly different in mass, yet so chemically and electro-chemically alike (isotopic) that by chemical manipulation it has not been possible to effect the slightest separation. Physical methods (as distinct from radio-active processes) of separation are now being tried, and only in one case, that of evaporation, has any promising result been obtained.‡

It will be seen that the term *isotope* (not to be confused with the term *isotrope*) stands for an important factor in analysing the atomic weights. This term means equal in place, the place being any particular one in the Periodic Table, thus signifying chemical equality, so that all the elements falling into a given place are chemically and electro-chemically alike. In accordance with the atomic-number idea every element of such a group-place would have the same atomic number. In studying the properties of the radio-active products which are definite atoms during their period of existence, it was discovered that certain members though from different radio-active sources (or those from the same source) could not be separated by chemical methods when those chemically alike were mixed together. It was also found that in many cases such like elements comprised atoms which had *different* atomic weights ranging over several whole-number units. Such elements

---

* *Phil. Mag.* (Dec. 1919), 38, p. 707.   See also *Camb. Phil. Soc. Proc.* (1920), 19, p. 317.

† W. D. Harkins was one of the first to use the term *species*, as applied to isotopes, which is one frequently used by the present writer.

‡ J. N. Brönsted and G. Hevesy—*Nature* (1920), 106, p. 144.   See also centrifugal experiments by J. Joly and J. H. J. Poole, *Phil. Mag.* (1920), 39, p. 372 ; also p. 376.

could, therefore, occupy the same place in the Periodic Table, and they were designated isotopes by Soddy who had, as is well known, made a special study of radio-active phenomena; together with Rutherford and many others, he contributed largely to that branch of chemistry known as radio-chemistry or radio-activity. In this connection mention must be made here of Becquerel and particularly of M. and Mme. P. Curie, whose researches on radium led to that branch of science known as *radio-chemistry*.* Returning to the subject of isotopy, it was found that the stable end-products of radio-active change from two different head or parent sources, uranium and thorium, were isotopic, these products being pure lead. One was a final product of uranium (there is a branch line in the descent, but this need not be considered here as it does not affect the principle involved), and the other a corresponding end-product of thorium. It has been possible to obtain these leads separately from different minerals, and very accurate determinations of their respective atomic weights has revealed a difference of mass of about 2 units, uranium lead being 206·08 and thorium lead 207·9. The theoretical values deduced from the Rutherford-Soddy disintegration law (see Appendix VI) should be 206 and 208 respectively. In citing experimental values some allowance has to be made for a small percentage of other leads being present. Ordinary lead not apparently of radio-active origin is (or appears to be) often absent and it seems to have been evolved independently of radio-active change as if it were a product of *evolution*, whereas the radio-leads (those of radio-active origin) are the product or result of *devolution*. The constancy of the atomic weight of ordinary lead, which has a different value from all others, is of interest. The following determinations of atomic weight are the most trustworthy ones :—

| | |
|---|---|
| Baxter and Grover (1915) . . . . | 207·20 |
| Richards and Wadsworth (1916) . . . | 207·183 |
| Richards and Hall (1917) . . . . | 207·187 |
| Average . . | 207·190 |

See in this connection Chapter XIV and Table XII (p. 102), also Appendix III.

All the leads discovered which are isotopic give barely any difference in their optical spectra. L. Aronberg † compared the spectrum of ordinary lead (strongest line) with that of lead of radio-active origin obtained from Australian carnotite (strongest line). He used a Michelson 10-inch plane grating in a Littrow mounting of 30 feet focus, with the two arc lamps employed connected in parallel to the vacuum pump to avoid any difference in pressure and thereby eliminate the possibility of any relative shift in the line-positions on the plates due to any pressure difference. Other conditions were

* See Rutherford, *Radio-Active Substances and their Radiations* (1912).
† *Astrophys. Journ.* (1918), 47, p. 96.

maintained the same. Measurement of the lines on 16 plates showed that the radio-lead gives relatively to the ordinary lead a line slightly longer in wave-length ; the average difference between the lines due to the two leads was 0·0044Å unit. When the ordinary lead was used in the second arc instead of the radio-lead, there was no difference in the lines obtained, thus showing that there could hardly have been any error in the measurements. The respective atomic weights of these leads were close to the values 207 and 206. It was concluded, moreover, that ordinary lead is not a mixture of thorium and radium lead, but is an isotope by itself. Of course, it may be here remarked that there is no evidence that all the atoms of ordinary lead are exactly alike in mass, but since there is a slight spectroscopic difference between the ordinary and the radio-lead, it may be assumed that an atomic weight difference exists between the two leads in question, and this is apparently the case. Prof. T. R. Merton,* by means of an improved method of measuring the interference fringes obtained by means of a Fabry and Perot interferometer, confirmed Aronberg's result, the accuracy of the method being of the order of 0·001Å unit. The following are the extended results of this later work :—

> Uranium lead †—Less refrangible than ordinary lead by 0·0050Å ± 0·0007.
> Ceylon-Thorite lead—More refrangible than ordinary lead by 0·0022Å ± 0·0008.
> Thallium from pitchblende—More refrangible than ordinary thallium by 0·0055Å ± 0·0010.

It would appear that thallium in pitchblende is an isotope of ordinary thallium.

There are other lines of attack, which are of interest, as instanced by the following : E. S. Imes ‡ in examining the absorption bands of HCl found certain doublets with components which F. W. Loomis § attributes to isotopes ; as, theoretically, if these bands are produced by the " vibration of the nuclei along the line joining their centres " in the molecule, it is shown that the frequencies should be approximately proportional to the square root of the effective mass $(m_1+m_2)/(m_1 \times m_2)$—where $m_1$ is the mass of the hydrogen nucleus and $m_2$ the mass of the chlorine atom. Therefore the band-lines due to H(Cl=35) and H(Cl=37) ‖ should differ by 1/1330, and those due to H(Br=79) and H(Br=81) should differ by 1/6478. In the case of the HCl band at 1·76 μ, Imes found doublets with components which are now shown to agree in separa-

\* *Roy. Soc. Proc.* (1920), 96, p. 388.
† Lead from pitchblende.
‡ *Astrophys. Journ.* (1919), 50, p. 251
§ *Nature* (1920), 106, p. 179 ; and *Astrophys. Journ.* (1920), 52, p. 248.
‖ See below and Table II.

tion and relative intensity with the theoretical results. In other cases Loomis points out that the computed separations are less than the resolving power of the instruments so far used.

The principle of isotopy is best illustrated by taking all the radio-active elements and tabulating them so that those of like chemical properties (experimentally determined) will fall together under one particular atomic number in each case, as shown by Table I.

TABLE I

| Group Number* . | III | IV | V | VI | O | II |
|---|---|---|---|---|---|---|
| Atomic Number . | 81 | 82 | 83 | 84 | 86 | 88 |
| Leading Element† | Thallium | Lead | Bismuth | Radium F‡ | Ra Emanation | Radium |
| | $RaC_2$<br>AcD<br>ThD | RaB<br>RaD<br>Ra-end<br>AcB<br>Ac-end<br>ThB<br>Th-end | RaC<br>RaE<br>AcC<br>ThC | RaA<br>$RaC^1$<br>AcA<br>ThA<br>$ThC^1$ | AcEm<br>ThEm | AcX<br>MsThI<br>ThX |

| Group Number . . . | III | IV | V | VI |
|---|---|---|---|---|
| Atomic Number . . . | 89 | 90 | 91 | 92 |
| Leading Element . . . | Actinium | Thorium | Uranium $X_2$ | Uranium (U). |
| | MsThII | Io<br>RdAc<br>RdTh<br>$Ux_1$ | | |

See also Tables XI (p. 101), XII (p. 102) and Fig. 61 (p. 192).

Returning now to Aston's recent experiments which extend the principle of isotopy to the ordinary elements and reveal the remarkable fact that they all have whole-number-atomic-weight atoms except atomic hydrogen which has, in a relative sense, an appreciable decimal fraction, the universal principle of isotopy is thus revealed. This principle and its extension to the common elements were first suggested by Soddy, although Crookes § many years ago gave expression to the possibility that the atomic weights were not relative numbers representing a definite weight of a definite atom but were *mean* values, the atoms of each element

* As ordinarily understood.
† See Appendix I for full names of the radio-elements.
‡ Or polonium.
§ *Chem. News* (1887), 45, p. 55.

differing appreciably in atomic weight.  Following along similar lines the writer * foreshadowed and developed a whole-number idea before the experimental evidence had accumulated ; in fact, there was at the time considerable evidence against such an hypo-thesis, as chlorine was not then regarded as being isotopic, since Thomson † had not found any lines (parabolas) on his plates which he had regarded as representing isotopes, except in the isolated case of neon, and this was held by some as not yet proved.  Aston now shows that chlorine is made up of atoms of masses 35·00 and 37·00 in about the proportion of 3 to 1 respectively, so that the mean value is 35·46, which is the experimental value obtained by chemical methods of analysis.  The isotopic values and the calcula-tion involved in this conception, which is a very simple one, can be shown conveniently thus :—

$$35 \times 3 = 105$$
$$37 \times 1 = \phantom{0}37$$
$$4 \quad )\overline{142}$$
$$35 \cdot 5$$

In all modern work oxygen is taken as 16 exactly, and this standard instead of H=1·000 is now amply justified by the atomic masses determined experimentally by Aston, which are given below in Table II.

At this juncture it is necessary to point out that the enormous amount of extraordinarily accurate atomic-weight determinations, by such experimentalists as T. W. Richards, of Harvard, and his co-workers, who have placed on record very exact magnitudes.  The work by various experimenters on chlorine alone, if bound into one set of volumes, would doubtless fill a small library.  Chlorine has an atomic mass very close to the above mean figure, and silver, for example, may be taken as 107·88 (O=16), the error in these cases probably not being great enough to alter the second decimal figures by more than one unit.  There are still many elements which have not been accurately measured, but the provisional atomic weights, using this term in the sense of close approximation, are given to the *first* decimal place only in such instances.  One may come across such statements as the following :—

" Analyses of various halogen compounds of silver give numbers ranging from 107·67 to 108·09 for the combining weight of silver (oxygen=16) ;  the best representative value is supposed to be 107·88." ‡

This statement simply means that historically the experimental range of values is *about* as given, but as the methods are improved

* *Chem. News* (1914), 109, p. 169.
† See Sir J. J. Thomson's statement in *Engineering* (1919), 106, p. 453.
‡ J. W. Mellor, *Modern Inorganic Chemistry* (1912), p. 381.

and refined, the value is narrowed down to a very small range, say, from 107·85 to 107·90 in this particular case. Of course, the skill * of the experimenter enters into the matter and it is only in the hands of a comparatively small number of experimentalists and their co-workers that thoroughly trustworthy results can be relied upon. Richards in America, Hönigschmid in Germany, and Aston in England, to name three distinguished contributors to this important department of chemistry, may be taken as leaders in atomic-weight determinations. There are a few others whose work takes equal rank with that of these experimentalists, but it is not necessary to extend the list here.

In Table II are shown the atomic weights of the atoms which have up to the present been accurately measured by the positive-ray method. All the values here given are those recently determined by Aston, †except in the case of magnesium, which was determined later by A. G. Dempster.‡

### TABLE II

Hydrogen . . . No isotopes ; atomic weight 1·008.
Helium . . . . No isotopes ; atomic weight 4·00.
Boron . . . . Two isotopes ; atomic weights 10 and 11.
Carbon . . . . No isotopes ; atomic weight 12 exactly.
Nitrogen . . . No isotopes ; atomic weight 14 exactly.
Oxygen . . . No isotopes ; atomic weight 16 exactly.
Fluorine . . . No isotopes ; atomic weight 19 exactly.
Neon . . . . Two isotopes ; atomic weights 20 and 22.§
Magnesium . . Three isotopes ; atomic weights 24, 25 and 26.
Silicon . . . . Atomic weights 28 and 29 (and 30 ?).
Sulphur . . . Atomic weight 32 (and one other ?).
Chlorine . . . Two isotopes ; atomic weights 35 and 37.
Argon . . . . Two isotopes ; atomic weights 36 and 40.
Arsenic and Phosphorus appear only to have atomic weights 75 and 31 respectively.
Bromine—Two isotopes ; atomic weights 79 and 81.
Krypton—This element appears to have as many as 6 isotopes of atomic weights, 78, 80, 82, 83, 84 and 86.
Xenon and Mercury are multi-isotopic, like Krypton, but the values, though apparently whole numbers, are as yet uncertain.
See concluding part of Chapter VIII, p. 53 ; also foot-notes on pages 48 and 147.

For a more complete list of isotopes, see Appendix I, pp. 191 and 193.

---

* Skill or technique is not an adequate word, since extensive knowledge of chemical reactions and the causes of error that are likely to arise become also of prime importance in work of this kind.
† See *Phil. Mag.* (1919), 38, p. 707 ; 39, pp. 449, 611 ; 40, p. 628.
‡ See *Science* (Dec. 10, 1920), vol. 52, p. 559.
§ First determined by J. J. Thomson.

NOTE.—See Annual Report of the International Committee on Atomic Weights for the latest *mean* values (should isotopes be involved) as the positive-ray method has not yet been so perfected as to indicate with any degree of accuracy the proportionate numbers of isotopic atoms.

From the foregoing it will be seen that *the principle of isotopy is found to be general throughout the Periodic Table*, that is to say, from lithium to uranium. The non-isotopic elements—those containing or comprising no atoms of appreciable difference in atomic weight—thus far discovered experimentally are given in Table II above, but there are comparatively few of these to be expected when all the isotopes are discovered, judging from the many atomic weights which depart appreciably from whole numbers. *The appreciable fractional irregularities of the atomic weights may be attributed to the existence of isotopes.*

Amongst the radio-atoms there are a few which differ chemically and spectroscopically from one another, yet they may have the same atomic weight when neglecting any small difference that might be due to the loss of an electron the mass of which is of the order of $0.00055$ relative to hydrogen taken as unity. Soddy and A. W. Stewart have suggested extending the use of the word *isobare* to such atoms. The isobaric principle may apply to some of the ordinary elements. For example, two types of elements may exist which are spectroscopically and chemically different yet both having atoms alike in mass. Should these exist they would not be isotopes by comparison but isobares (see Table XIX, pp. 148, 149). These might be described as heterotopic isobares, as distinguished from isobaric isotopes, to use Soddy's terminology.

Referring now to those elements not consisting of isotopic atoms in the general application of the term as, for example, oxygen and carbon whose respective atoms are all of the same mass, neglecting any remote decimal-place fractions, they could, according to the above nomenclature, become isobaric isotopes, as they would have the same weight and occupy the same periodic place.

It is now evident, as Soddy * has pointed out, that the lack of mathematical relationships between the atomic weights, as hitherto understood, may be explained on the assumption of isotopes. Moreover, now that the atomic weights are all turning out to be whole numbers, the problem of atomic constitution may be much simplified.

It would appear that the diversity of matter is not due to a multiplicity of elements whose atomic weights differ in a complex way so much as to the manner in which simple units common to all elements can give rise to varying force distributions. These units

* *Chem. News* (1913), 108, p. 168.

appear to be electrons and atomic nuclei (see next Chapter). The electrons appear to be all alike, and the nuclei may be the same when considered in the terms of their fundamental units. The combinations are, however, different, representing " a mighty maze but not without a plan."

## REFERENCE

F. W. Aston, *Isotopes* (1922).

# CHAPTER III

## ELECTRONS AND ATOMIC NUCLEI:
### SUB-ATOMIC PHENOMENA

VACUUM-TUBE phenomena have led to the elucidation of the properties of the electron, one of Sir J. J. Thomson's great achievements, but the preliminary work of J. S. Townsend must be regarded as paving the way. C. T. R. Wilson and E. Wiechert have also contributed to the knowledge of the electron ; there are other workers too, but space will not permit a reference to them here ; suffice it to say that Wilson's experiments showed that ions could serve as condensation nuclei for water vapour in the formation of droplets or clouds, the ions being produced by rays from radio-active matter or X-rays. The cloud was formed by the sudden expansion of the gas, the cooling effect causing the water vapour to condense in cloud-forming drops round the nuclei produced by the ionisation. Photographs were obtained of ionised tracks in gases which were rendered visible by the deposition of water on the ions. The trajectory of an invisible particle passing through moist air could thus be rendered visible by the droplets left suspended in its track. An $\alpha$-particle, a helium atom or nuclear part given out explosively in the process of radio-active change, can be rendered, as it were, not only visible by the aftermath-track of droplets, but when the particle encounters an atom or molecule in the gas through which it is passing, its deflection can also be recorded by these condensation streaks of cloud, which they really are. Consequently the tracks sometimes show deviations and spurs, thus :—

FIG. 3.

The arrow-head shows the direction of movement of the particle.

By the word *ions* one usually understands atoms or molecules in the abnormal state of being electrically charged either by the capture or retention of negative electrons or by losing electrons, so as to leave them with a charge or field. In some cases the atom may lose several electrons, in which event its positive charge would be correspondingly increased.   Free negative electrons might, of course, function as ions in so far as they represent negative charges of electricity, but the term *ion* is usually associated with an atom or molecule.   A droplet of oil would be a cluster of molecules. Such a cluster carrying one or more electrons would not constitute an ion in the orthodox sense, as exemplified in the $NO_3^-$ ion (see second paragraph below).

Mention must be made that Prof. R. A. Millikan [*] has succeeded in capturing ions on droplets of oil (mercury was also employed) and, by observing them through a low-power telescope with spaced hairs, their up-and-down movements in an adjustable electric field were studied.   The charged droplets were held in the field of view against gravity by adjusting the opposing field strength.   The fall and rise of the droplets were then accurately timed in transit across the hairs. . . .   In short, the droplets could be electrically weighed, knowing the field-strength, and their times of fall or rise, by making use of a modified Stokes' law of falling drops.   *A method of practically direct measurement of great accuracy was thus developed and the charge of the electron in absolute electrostatic units was found always to be a multiple of* $4.774 \times 10^{-10}$*, with an accuracy estimated at one part in a thousand.*   Measurements and computations agreeing closely with this value have been made from time to time by other methods, as shown substantially by Table III, as compiled by J. A. Crowther,[†] the cases being selected ones.

### TABLE III

CHARGE ON AN ELECTRON IN ELECTROSTATIC UNITS

| Observer | | | Method | Values × $10^{-10}$ |
|---|---|---|---|---|
| Begeman | . | . | Water cloud (Wilson) | 4·67 |
| Millikan | . | . | Oil and mercury droplets | 4·774 |
| Perrin . | . | . | Brownian movements | 4·2 |
| Rutherford . | . | . | Charge on α-particle | 4·65 |
| Regener | . | . | Charge on α-particle | 4·79 |
| Planck . | . | . | Theory of radiation | 4·69 |

Prof. Millikan and his co-workers [‡] have shown that $\alpha$, $\beta$, $\gamma$ and X-ray ionisation consists, in at least 99 times out of 100, in the detachment in each case of single negative electrons from the neutral atom.   Many gases and vapours were tested by the droplet method and no authentic case appears where more than one electron was detached.   The electron, after being severed from its atom by the direct action of the rays, leaves an ionised molecule behind which is caught on the oil-droplet and electrically ' weighed.'

[*] See *The Electron* (1917).
[†] *Ions, Electrons and Ionising Radiations* (1919).
[‡] *Am. Nat. Acad. Sci. Proc.* (1919), 5, p. 591 ; *Phys. Review* (1920), 15, p. 157.

There is thus a rather clear demonstration of ionisation of a non-multivalent type as measured by the displacement from the atom of a single electron. These rays, including $\alpha$-rays, are supposed to have passed practically through the atom. The term *valent* is borrowed from chemistry. Atoms combine in definite proportions owing to electrical conditions which have received numerical representation in terms of the number of, say, hydrogen atoms that will combine with a given atom, hydrogen being taken as unity. Oxygen has a value *two* assigned to it, since it will combine to form the water-vapour molecule, which is $H_2O$. It is said to be bivalent. This phase of the subject will be detailed further on in the light of recent work (see Chapters XIII, XIV and XV).

M. J. Kelly * has studied the photo-emission (photo-electric effect : see Chapter XI) when ultra-violet light is allowed to fall on droplets of sulphur, shellac, oil and paraffin, with the result that in each case only one electron escapes at each emission. This work and the results recorded are similar to that given above by Millikan and his students and confirms the valence magnitude associated with the action.

Passing now to discharges in vacuum-tubes, the following, taken from one of Sir J. J. Thomson's expositions of the subject † under elucidation, should be of interest : " We have in the course of the last fifteen years attained to very definite ideas as to the nature of negative electricity. Negative electrification has been shown to be due to the presence of minute particles called corpuscles or electrons, all of which are exactly the same kind—that is, each particle carries the same quantity of negative electricity, and each has the same mass. This mass is far smaller than that associated with ordinary matter, being, unless the velocity of the corpuscle is comparable with that of light, only about $\frac{1}{1700}$ [later estimate $\frac{1}{1840}$] of the smallest mass hitherto known to science, that of an atom of hydrogen. The mass of these corpuscles increases rapidly when their velocity approaches that of light, and if their speed were about 50 yards per second less than that of light, their mass would be about the same as an atom of hydrogen. These corpuscles may exist in a free state, as in the cathode rays in a vacuum-tube, or they may be attached to atoms, or molecules, as when a solid or liquid body is negatively electrified. They form a portion of the atoms of all kinds of matter, each atom containing a definite number of corpuscles. For atoms other than hydrogen the number of corpuscles in the atom is about half the atomic weight ; thus helium, whose atomic weight is four, has two corpuscles in its atom ; oxygen, whose atomic weight is sixteen, has eight ; and so on. Hydrogen, whose atomic weight is one, has probably one corpuscle in its atom. The process of electrification consists in taking corpuscles from one body and giving them to another. On this view,

---

* *Phys. Review* (1920), 16, p. 260.
† *Harper's Monthly Magazine* (1914).

negative electricity is molecular in structure, and all negative charges are integral multiples of a certain unit whose value is now known with great accuracy [see Table III]. In fact, we may regard Franklin's electric fluid as a collection of such corpuscles, only we must suppose that an excess of this fluid corresponds to negative electrification, and not to positive as Franklin supposed.''

The nature of electrical phenomena makes it necessary to postulate a positive electron, but this electron has not been definitely isolated. Rutherford seems of the opinion that this unit is hydrogen itself taken apart from that of the negative electron normally associated with it, and it thus becomes the positive counterpart which gives the atom over 99·9 per cent of its mass. See in this connection Chapter XX.

The negative electron has a mass of $\frac{1}{1840}$th part of the hydrogen atom of atomic weight 1·008 ($0 = 16$ exactly); or expressing the mass of the hydrogen atom in gramme-units, it is $1·64 \times 10^{-21}$, whilst similarly the mass of the electron is $8·9 \times 10^{-28}$ gramme; so that

$$\frac{1·64 \times 10^{-24}}{8·9 \times 10^{-28}} = 1840.$$

Now the charge on the electron—or of the electron—is $1·57 \times 10^{-20}$ in electromagnetic units. This is usually symbolised by the letter $e$. The charge-to-mass ratio, $e/m$, for the electron is $1·774 \times 10^7$ in electromagnetic units per gramme, $m$ being the *electromagnetic mass* of the electron.

Whether the more massive part of the atom should be taken as purely electric is not a settled point. Perhaps it should, but it cannot be regarded in every sense the mirror image of the negative electron since its properties appear to be somewhat different from those associated with the latter.

The process of radio-active change gives a very complete picture of atomic disintegration, and by studying successive changes in the structural composition of the atom, revealed with remarkable clearness in the case of the radio-atoms, certain mutually-supporting facts are brought to light—especially when these are coupled with those revealed by the vacuum-tube.

In those cases where the electron may be regarded as moving with a velocity below $\frac{1}{10}$th that of light its mass is $\frac{2}{3}\mu \times e^2/r = 8·9 \times 10^{-28}$ grm., in which $e$ is the electric charge of the electron, $\mu$ the permeability, that of a vacuum being taken as unity, and $r$ the radius of the charge taken as a sphere. This formula is derived from electromagnetic equations and it has been apparently verified. Therefore, knowing $e$, which is $1·57 \times 10^{-20}$ in electromagnetic units (or $4·7 \times 10^{-10}$ in electrostatic units), the radius $r$ can be deduced, and this is about $1·8 \times 10^{-13}$ cm. The radius of an atom of helium is about $1 \times 10^{-8}$ cm., hence with respect to this atom *as a whole* the electron is very small : the ratio being about 1 to 55,000.

An electron expelled from a radio-atom is, of course, a β-ray. The β-ray speeds vary in different radio-atomic changes. Some of these rays travel with a speed of about $\frac{1}{10}$th that of light—that is to say, they are ejected from the atom at this speed. It seems necessary to assume that the velocity of the electrons in the normal atoms (those not disintegrating) cannot much exceed this speed, for at higher speeds, the highest being $\frac{9}{10}$ths that of light, the mass of the electron increases ; but in such cases the " transverse mass " has to be considered (see Appendix II).

An α-particle colliding with an atom of gold of nuclear charge 79 units will be turned back in its path at a distance of $3 \times 10^{-12}$ cm. from the atomic centre, indicating that the nucleus of the gold atom has a powerful electrostatic field concentrated as a point charge. This is an experimental fact. J. Chadwick * has measured, by a method based on the scattering of the α-particles in passing through thin metal foils, the charges on the nuclei of platinum, silver and copper, these values being 77·4 (78), 46·3 (47) and 29·3 (29) respectively, the already-established atomic numbers being shown in brackets.† The inverse square law holds good accurately in the region round the nucleus up to within $10^{-11}$ cm. of its centre. No electrons are believed to be present between the nucleus and the K-ring. The K-ring is the one which is supposed to give rise to the K-radiation (see Chapter V).

The foregoing small figures are not much greater than the radius of the electron itself, so that one may assume that the formula derived from electromagnetic equations may be applied to the positive nuclei. Therefore, since the mass of the hydrogen atom is $1·64 \times 10^{-24}$ and its positive charge is equal to that of its electron, hydrogen having only one electron, viz. $1·57 \times 10^{-20}$ electromagnetic unit, the radius becomes

$$\frac{(1·57 \times 10^{-20})^2}{1·64 \times 10^{-24}} \times \tfrac{2}{3} = 1 \times 10^{-16} \text{ cm.}$$

It will, therefore, be seen that the positive electrification in the case of hydrogen is concentrated in (or on) a sphere $\frac{1}{1800}$th that of the electron in size ; and, as probably the figure for the gold atom is an upper limit, the nuclei of all atoms have radii ranging from

$$10^{-16} \text{ to } 10^{-12} \text{ cm.}$$

This is not a statement based entirely upon pure theory, for experimental data give support to it in a most direct way.

These values when considered comparatively, together with the possible revolution of the electron, seem to show that the atom is like a planetary system, but of inconceivable smallness.

The equation here used is indirectly supported by the work of Kaufmann (1901) and others, who found that the deflection of β-rays from a radium product could be accurately photographed as a curved line, the method being similar to that of the positive-

* *Phil. Mag.* (1920), 40, p. 734.      † See Chapters IV, V and VIII.

rays described above. Had the line been a straight one, showing no acceleration, the mass of the particle would *not* have varied with the speed.

Having arrived at some idea of nuclear dimensions, the problem of visualising multiple nuclei presents itself, since it becomes necessary to consider the atom as a complex structure, especially when studying the heavier elements. Organic chemistry is suggestive in this connection. It has been pointed out by shrewd observers that there should not be a fundamental distinction between organic and inorganic chemistry; and no doubt the working separation of the two branches has given rise to a sort of feeling that organic chemistry represents a group of phenomena having no counterpart in other phenomena, whereas there is no sharp line of demarcation. That is to say, every phenomenon has its counterpart, and now it is being discovered that the atomic structures simulate molecular structures. One should, therefore, look closely into organic chemistry for analogous clues in studying the atomic structure. One may, for instance, look for ring structures and possibly the atoms of the inert gases are really closed rings of sub-atoms without stray fields, or without a weakness in the chain or ring section, so that it cannot be broken or linked up with other atoms or those of its own kind. One can begin to see why helium liberated from a radio-atom structure has two positive charges. It may be an open link de-linked from a chain or ring formation. This view is, however, too uncertain to be seriously considered. One can call to mind the charges of the inert-gas atoms unmistakably shown by positive-ray analysis, and these could hardly represent broken rings, or any equally drastic alteration in the atomic structure.

In addition to electrons being expelled from a disintegrating atom, $\alpha$-particles (atoms of helium) are expelled, but rarely the two together, as each atomic step is accompanied by a definite type of 'single' emission showing that the atomic edifice contains bricks (helium units) and binding electrons, the latter possibly taking up a more stable position in the atom when the former is expelled. Hydrogen particles do not appear to be expelled. They are probably present, however, as part at least of the helium unit which is very stable. This unit may be evaluated thus (see Chapters VIII and XX)—

$$He = 1+1+1+1, \text{ or}$$
$$He = 1+2+1, \text{ or}$$
$$He = 2+2, \text{ or}$$
$$He = 1+3.$$

Probably "3" would be equal to $H_3$ * and "2" equal to $H_2$, but see particularly Chapter XX. These ideas are now being in-

* $H_3$ has apparently been prepared by the activation of hydrogen. See G. L. Wendt and R. S. Landauer, *Am. Chem. Soc. Journ.* (1920), 42, p. 930; see Index.

vestigated experimentally by Rutherford, but just what the detailed outcome will be it is impossible to say. The disruption of the nitrogen atom by Rutherford, and later the disruption of the oxygen atom, gives strong support—on top of that of radio-active phenomena—to the idea that all atoms are made up from simple positive units of mass values, possibly 1, 2, 3 and 4, together with binding electrons (see Chapter VIII).

The following gives perhaps expression to the ideas more or less current at the present time : " Radio-active phenomena indicate that possibly helium is a common constituent of all elements except hydrogen, in which case its combination must be of a superior order to anything coming within the range of controllable phenomena such as chemical action, but like the molecule in crystals the definite individuality of such a fundamental helium atom or unit taken as a sub-atom may be lost, or merged into a larger whole. In this case a deep-seated type of valence may exist accompanied with an affinity far superior to anything exhibited in chemical phenomena, just as ordinary valence is in a sense more fundamental and definite than that which is connected with the formation of molecular complexes. The gradation is fairly complete as the last-named phenomenon shades off into cohesion. It will be seen that this line of reasoning leads to four physico-chemical steps, namely :

" 1. Helium atoms as sub-atomic units involving a superior type of valence and affinity. A further subdivision into similar hydrogen units has been foreshadowed as a possibility, in which case a similar antecedent step to this one might be expected.
" 2. Atomic combinations under control of the chemist involving ordinary valence, but with affinities which are not co-related to the valence values.
" 3. Molecular complexes, involving a sort of residual or excess valence which tends to become diffuse.
" 4. Cohesion, involving a still more remote type of valence which being wholly diffuse is more of the nature of an affinity pure and simple, and, indeed, in this case the idea of *valence* may be entirely eliminated."

The foregoing statements * are of a general nature. It will be seen from the octet theory (Chapters XIII, XIV and XV) how definite ideas take the place of the more general ones. One is reminded, however, of the different methods of theoretical attack employed. As an example, N. R. Campbell distinguishes between *mechanical* and *mathematical theories*, as the following quotation from his recent book † will show : " Our inquiry thus narrows

* Taken from an article by the writer, *Chem. News* (1919), 118, p. 145.
† *Physics : The Elements* (1920), p. 152.

down to the question whether, by reason of their nature, theories of the first type [mechanical] are more liable to error than those of the second. It may be answered at once in the affirmative. The making of any additional statement is necessarily accompanied by an additional possibility of error, and theories of the first type are more liable to error than those of the second simply because they state more. Theories of the second type can only state relations between laws of the same kind and involving the same concepts ; theories of the first kind state relations between laws of different kinds involving different concepts."

NOTE.—In referring to the $\alpha$-particle—a helium atom having two positive charges—in many places in the text the fact of its charge will not be mentioned. It is understood that this particle is differentiable from the ordinary neutral helium atom by reason of this double charge. It is also referred to as a nuclear unit. When ejected from a radio-active atom the term $\alpha$-ray is frequently used.

# CHAPTER IV

## CO-ORDINATING THE ELECTRICAL PROPERTIES OF THE ATOM

THE periodic classification of the radio-active elements affords perhaps the most convincing illustration of the electric property of the atom and its structure. This may be referred to as the Russell-Fajans-Soddy classification * owing to the contributions of these workers in giving expression to the experiments of Fleck, G. von Hevesy and others. The table in Appendix I is an extension of the standard type of table as first imperfectly expressed by de Chancourtois in 1862, then by Newlands in 1863-64, and later perfected by Mendeléeff in 1869 — to mention the leading contributors to the generalisation ; but perhaps Lothar Meyer deserves special mention since in 1860 he observed that " if the elements be arranged in the order of their atomic weights there is a regular and continuous change of valence as we pass from family to family, and that the successive differences of elements in the same column are at first approximately 16, except beryllium, then they increase to about 46 and finally approach a number ranging from 87 to 90."

In the table of the radio-elements (see Appendix I) it is shown how they may be systematically classified into groups in accordance with the *law of change*, which may be stated thus : When the atoms of a given element lose helium atoms (one from each) it becomes another *type* of element, fitting into a place two group-units less in the Periodic Table. The helium atom given off may be regarded as a sub-atom when forming an integral part of the original atom. It is shot out as an $\alpha$-particle and electrically possesses a double positive charge. This might be illustrated by saying that element $A$ of, say, Group IV spontaneously ejects a helium atom of mass 4 and carrying two positive charges, $A$ then becoming element $B$ which fits into Group II and has an atomic weight 4 units less. When the atoms eject electrons (one from each) which are shot out as $\beta$-particles (rays), this being another name for electrons when they are being ejected from the atomic system, the element becomes one of another type and passes to a group-place *one* unit higher up,

---

* A. S. Russell, *Chem. News* (1913), 107, p. 49 ; K. Fajans, *Phys. Zeit.* (1913), 14, pp. 131, 136 ; Soddy, *Chem. News* (1913), 107, p. 97 ; also 108, p. 168

e.g. passes from IV to V. In this case the atomic weight does not alter appreciably (the mass of the electron being about $\frac{1}{1840}$th of that of the hydrogen atom—see Chapter III); and this fact was one of the clues that led to the idea of isotopes : showing that *the mass or weight of an atom does not in itself determine its chemical difference.* Since each atom when it suffers this change loses one negative electron, this is electrically equivalent to *gaining* one positive charge, hence the backward step (see Appendix I).

These changes in the electrical state of the atom fit in exactly with the idea first put forward by van den Broek : " That the scattering of $\alpha$-particles by the atoms was not inconsistent with the possibility that the charge on the nucleus was equal to the atomic number of the atom, i.e. to the number of the atom when arranged in the order of increasing atomic weight," and this order may be established more in detail when the isotopic system is taken into account, an attempt in this direction having been made (see Chapter XIX). It is to Moseley, however, that the credit is due of having greatly strengthened the atomic-number theory as applied fundamentally to the atom. " Moseley showed that the frequency of vibration of corresponding lines in the X-ray spectra of the elements depended on the square of the number which varied by unity in successive elements. This relation received an interpretation by supposing that the nuclear charge varied by unity in passing from atom to atom [the term *atom* here including its isotopes in cases where they exist] and was given numerically by the atomic number." The atomic number is therefore believed to be the nuclear charge, and it is considered of fundamental importance. The recent experiments of Chadwick, noted in Chapter III, afford further evidence as to the soundness of the atomic-number conception. The proportionality thus discovered by Moseley is not a direct one, as the frequency is proportional to $(N-a)^2$, where $a$ is a constant and $N$ is the atomic number. $a$ has one value for the K-series of lines and another value for the L-series (see Chapter V). It is supposed that the value of $a$ depends upon the nearness of the electrons to the nucleus and the number of electrons involved. The quoted passages in this paragraph are from Rutherford's Bakerian Lecture (see Chapter VIII).

The importance of the atomic number can be simply illustrated by a diagram (Fig. 4), which shows practically the spectral-line places, here represented as dots. P, Q, R, etc., will represent consecutive elements taken from that part of the series where the numerical order has to be changed in respect of two adjacent elements to bring those represented into chemical order, which is the order here given, thus showing that the atomic numbers agree in such cases with the order in which the element should occur to take up their right places in the Periodic Table. It must be remembered that each element may be made up of isotopic atoms (with some exceptions, as C, O, N and F) which make no difference in the X-ray spectra, so far as can be ascertained, and this is an important point to bear in mind. The high frequency spectra (X-ray spectra) of the lead isotopes are in absolute agreement there being no discernible difference, as should be the case since the atomic number (82) is the same for each isotope.

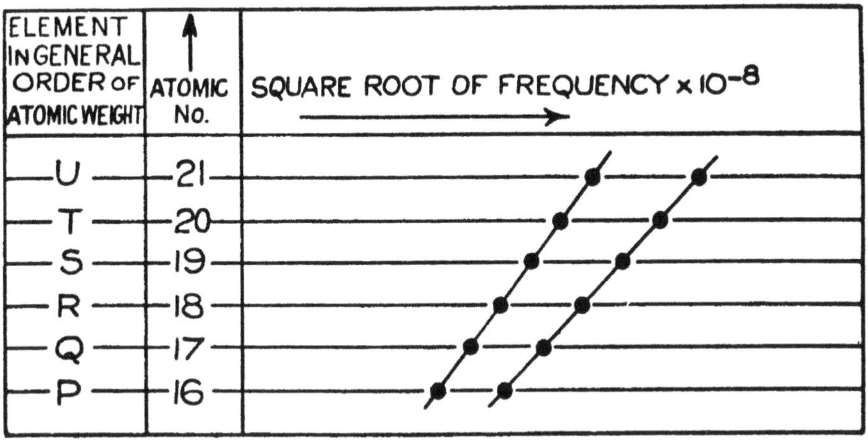

See Fig. 7 (after Bragg).

FIG. 4.

The atomic numbers start with hydrogen as *unity*, helium being No. 2, lithium No. 3, etc., up to uranium No. 92.

A brief mention of H. G. J. Moseley's procedure and a reference to the pioneer workers are of interest. W. H. Bragg and W. L. Bragg, following up the mathematical work of M. Laue and the experiments of Laue, P. Knipping and W. Friedrich, invented the X-ray spectrometer and determined the wave-lengths of the X-rays of a few metals. Following on this lead, Moseley carried out a series of elaborate determinations in which the cathode-rays (electrons in flight) were directed in successive experiments against anti-cathodes, each one of which was composed of a different element, except in a few cases when they were of different compounds—when the elements were not available or workable by themselves in the experiment. These anti-cathodes were mounted on a travelling carriage situated within the vacuum-tube, so that each element, or compound, could emit its characteristic radiation when brought into the range of the cathode stream. The movement of the carriage was under external magnetic control, an armature or ironpiece being fixed to the carriage. The emergent ray or pencil of X-rays was passed through an analyser consisting of a crystal of potassium ferro-cyanide, the rays being directed against the cleavage planes of the crystal which was set at the proper angle. At each setting of the carriage the diffracted rays thus produced were allowed to fall on a photographic plate. Upon developing the plate, or plates, the characteristic lines of the elements, successively forming the anti-cathode, were shown. This method of analysis has an advantage over the ordinary spectrum method, as the X-ray lines are few in number, being characterised by a pair of strong lines, which occupy regularly advanced positions in passing from element to element, as shown diagrammatically by Fig. 4 on previous page. The actual spectrum lines obtained from a set of elements from Ti to Cu are represented by Fig. 5. The lines on the right-hand side are a little stronger than those on the left. This step-wise regularity is a very important one and it has been extended to include practically all the elements. See Fig. 7 in the next chapter.

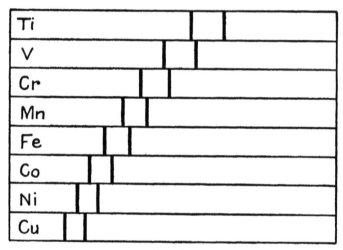

FIG. 5.

It is of interest to note that the lines are not influenced by the elements either being mixed together as in alloys, or combined chemically. For example, the lines of copper and zinc show clearly in the spectrum of brass given in Fig. 6.

FIG. 6.

These lines, besides establishing the element in a series, give strong support to the importance of the *atomic number*, which number seems to represent the *electric charge of the nucleus, or the number of electrons in the atom*. Comparison of the lines (pairs) of consecutive elements, arranged in the general order of increasing atomic weight, shows the presence of gaps in the orderly steps and indicates presumably that the elements answering to the line-pairs for the gaps have not yet been discovered. There appear to be six missing elements—one preceding ruthenium, one preceding samarium, one preceding tantalum,* one preceding osmium, one preceding radium emanation, and one preceding radium, the respective atomic numbers for these missing elements being 43, 61, 72, 75, 85 and 87. Those of atomic numbers 85 and 87 are probably radio-active and, of course, they may be extinct ; but there is some difficulty in linking up such elements with those of the known radio-active series. It is not safe even to predict the almost obvious, so that the filling of these gaps may not be accomplished altogether, as is generally anticipated. This comment is not intended to be an opinion, but merely a note of caution. See in this connection Chapter XX.

It is significant to observe that while the bombardment of the elementary substances in the solid form with cathode rays gives rise to an X-ray radiation yielding a characteristic spectrum, some of the radio-active elements (atoms) give out this very identical type of ' X-radiation,' which is designated $\gamma$-radiation ($\gamma$-rays) ; and, moreover, some of the same class of elements give out $\beta$-rays which are fast-moving electrons. These atomic radiations are given out when the atomic system partly disintegrates (all the radio-atoms only partly disintegrate), giving rise to a new type of element at each change. The third type of atomic radiation is that of the $\alpha$-rays which are helium atoms or nuclei of double positive charge, as already explained. This radiation is given out when the element undergoes change into one of two-group places (or two valences) less in its position in the series. Here again, by a reverse action, that is, by bombarding solid substances with these ejected particles, or in particular with $\alpha$-rays, a new phenomenon is brought to light which affords a key to unlock some of the secrets of atomic structure.

Geiger and Marsden have explored the regions of the innermost parts of atoms in thin films of matter by studying the trajectory of $\alpha$-rays as they enter and emerge from the film, with the result that in observing the scattering of these particles in their passage a small percentage underwent a deflection through an angle exceeding 90 degrees. This action was adequately accounted for by Rutherford on the assumption that each highly-deflected $\alpha$-particle had passed very close to the nucleus of a single atom in passing through the film, in fact, close enough to suffer a strong repulsive effect of the intense positive charge of the nucleus represented by the atomic number. It should be noted that the $\alpha$-particle itself carries a double positive charge, and, as ' like repels like,' as shown in all electrical phenomena, this repulsion is to be expected when the $\alpha$-particle passes close to the positive nucleus of the atom. The interpretation of this action therefore involves the idea of the *nuclear charge* being confined to a very small space at the centre of the atom, which, for a radial distance up to $3 \times 10^{-12}$ cm., behaves as if it were concentrated at a point. The point conception differs

---

* Langmuir assigns lutecium to this place (see Table XI), but introduces $Tm_2$ earlier in the series. See Appendix I (Notes).

from that of J. J. Thomson, who originally suggested that the positive mass was spread over the whole atomic sphere of influence, the electrons being embedded in it whilst they were in rapid orbital movement round the centre of the system. This idea as well as those along similar lines suggested by Lord Kelvin and by H. Nagaoka were highly instructive and represent the preliminary work that has prepared the way for the later developments.

Geiger and Marsden found that the number of particles scattered and the observed angle of scattering fulfilled the theoretical requirements (on the assumption that the forces between the $\alpha$-particles and the nucleus varied according to an inverse square law of distance) within the limits of experimental error over a large measured range. C. G. Darwin showed that the inverse-square law must apply in this case. Rutherford's conception of the nuclear constitution of the atom thus received very strong support from the scattering of $\alpha$-particles or rays, the atomic charge (atomic number) and the general trend of all the foregoing arguments and experiments, each one of which is of electrical character. It would seem that when matter is probed and the smallest of its parts revealed it is composed of truly electrical entities (see Chapter VIII).

# CHAPTER V

## THE K AND L SERIES OF LINES BY X-RAY ANALYSIS

.Dr. Kaye in his book *X-Rays* (1918) gives a short account of the characteristic X-radiations (see pages 116 and 211). Crowther also gives an account of these radiations in his book, *Ions, Electrons and Ionising Radiations* (1919) in sections 90 to 95. See also *X-Rays and Crystal Structure* (1918) by W. H. and W. L. Bragg.

A very brief account of this subject will be given here. When cathode rays, which are streams of fast-moving electrons emerging normally from the cathode, strike a metal like platinum, acting as an anti-cathode, a characteristic radiation is emitted that constitutes X-rays, these being of exceedingly short wave-length. Under suitable conditions the wave-length of these rays from platinum has been found to have the length $1 \cdot 10 \times 10^{-8}$ cm.*

### TABLE IV

| Kind of Wave | | | | | Wave-length in Cms. |
|---|---|---|---|---|---|
| Hertzian waves | . | . | . | . | $10^6$  to $0 \cdot 2$ |
| Infra-red rays . | . | . | . | . | $0 \cdot 031$  to $7 \cdot 7 \times 10^{-5}$ |
| Visible light rays | . | . | . | . | $7 \cdot 7 \times 10^{-5}$ to $3 \cdot 6 \times 10^{-5}$ |
| Ultra-violet rays | . | . | . | . | $3 \cdot 6 \times 10^{-5}$ to $5 \cdot 0 \times 10^{-6}$ |
| X-rays | . | . | . | . | $1 \cdot 2 \times 10^{-7}$ to $1 \cdot 7 \times 10^{-9}$ |
| $\gamma$-rays | . | . | . | . | $1 \cdot 4 \times 10^{-8}$ to $1 \cdot 0 \times 10^{-10}$ |

The apparatus used to measure these exceedingly short waves, known as the X-ray spectrometer, has become a very powerful tool for analysis. In Moseley's experiments the anti-cathodes were different elements, as described in the previous Chapter. The X-radiation is characteristic of the element emitting it, and it is particularised by two distinct lines (see Fig. 5). The relation between the frequency of these waves and the atomic number assigned to the element has already been indicated, the frequency varying as the square of the atomic number.

The penetrating character of these radiations increases as the

*¦Kaye gives the following table of wave-lengths of various radiations which are electromagnetic waves according to the accepted and well-proved theory of radiation. For theory of X-rays see page 72.

atomic weight of the element emitting them, and by using aluminium as an absorbing medium the penetrability (mass-absorption) can be plotted against atomic weight, giving two similar curves—one for the K-radiations and one for the L-radiations. For each metal the K-radiation (hard) is about 300 times more penetrating than the corresponding L-radiation (soft). These radiations are of different wave-length for different elements and they are often referred to by the name of the element giving the radiation ; thus silver, which gives a wave-length of $0{\cdot}560 \times 10^{-8}$ (K) and $4{\cdot}17 \times 10^{-8}$ (L), has a ' silver radiation ' of $0{\cdot}560 \times 10^{-8}$ when referring to the K-radiation. The platinum radiation is given above.

When the X-ray spectrometer is fitted with an analyser consisting of a reflecting crystal, which becomes to X-rays a diffraction grating (see Chapter VI), a spectrum of the radiation is produced which resembles the ordinary spectrum and the characteristic radiations appear as K and L series lines, but each series is made up of a strong line and a weaker one a little distance from it, as shown by Fig. 5. There are other fainter lines which appear on the photographic plates. These main lines, strictly speaking, may not be homogeneous but may consist of two or more closely associated lines which appear as one until resolved.*

These characteristic radiations are not limited to the elements emitting X-rays under bombardment with cathode rays (fast-moving electrons), but the radio-active emission of $\gamma$-rays ($=$X-rays also of atomic origin but spontaneously induced by the radio-active process) gives a series of values that seem to correspond with them. Radium C, for example, has a penetrating value of $0{\cdot}0424$ in aluminium and falls on the curve for the K-radiation when the atomic weight is plotted against the quality of the radiation corresponding with its penetrability in aluminium. With radium B there are three characteristic radiations, the second one of which corresponds to an atomic weight of 214, which is the right value for this radio-atom (see Appendix I). Ionium of atomic weight 230 gives an L-radiation which falls into proper line with the others when plotted in the manner indicated.

When the radiations are analysed by the spectrometer, as used by Moseley, the K and L series do not form a continuous set of regularly displaced lines from lithium to gold, but two sets appear which overlap from zirconium to silver ; in both cases the frequency of the radiation as registered on the photographic plate is proportional to the atomic number.

The foregoing statements will be made clearer from Fig. 7,

* M. de Broglie (*Comptes Rendus* (1920), 170, p. 1245) observes that since there is a general similarity between the lines of different elements, as instanced by the fine lines of helium and hydrogen, this principle may be extended to the X-ray spectra. Upon investigation of the K spectrum of tungsten, it is found that there is a doublet which has the wave-length difference of $0{\cdot}0007$ unit. Rhodium gives a doublet difference of $0{\cdot}0006\text{Å}$ unit, which points to the principle being general. See Chapter X : part dealing with Sommerfeld's work.

which shows the lines recorded ; but for the appearance of the principal lines see Fig. 5 in the previous chapter.

Dr. Crowther (*loc. cit.* p. 167) says : " The present limits of the two series [see Fig. 7] are due to experimental difficulties and each will no doubt be extended in both directions.  The K series for

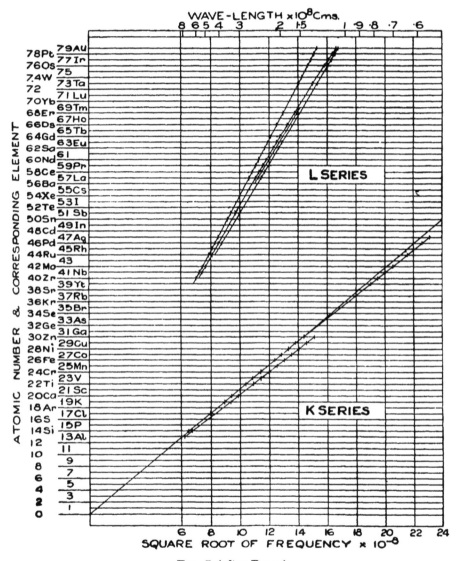

Fig. 7 (after Bragg).

elements of higher atomic weight than silver is difficult to excite . . . because the velocity of the cathode rays necessary to excite a given radiation increases with the frequency of the radiation. On the other hand the L-radiation from elements of low atomic weight is difficult to observe on account of its great absorbability.

" In addition to the K and L series there is evidence of at least

3

one, and probably more, similar series. The X-ray spectrum is at present very far from having been completely mapped.

"Since the K and L series correspond to lines in an optical spectrum we should expect some relation between them. It has been pointed out that if an element of atomic weight $A_K$ emits K-radiation of the same wave-length as the L-radiation emitted by an element of atomic weight $A_L$ then—

$$A_K = \tfrac{1}{2}(A_L - 48).$$

If this relation is universally true it follows that no element of atomic weight less than 48 can give out radiation belonging to the L series."

Dr. Kaye (*loc. cit.* p. 224) says: "Moseley . . . examined photographically the X-ray spectra of a large number of elements and obtained remarkable and important results. The elements were mounted as anti-cathodes . . . in an X-ray tube, the X-rays being analysed by means of a crystal of potassium ferrocyanide. The discharge tube was provided with an aluminium window (0·0022 cm. thick), which in those cases where the radiation was very soft was replaced by one of goldbeater's skin. In some instances the whole spectrometer had to be enclosed in an evacuated box, since the rays were too soft to penetrate more than 1 or 2 cms. of air."

It will be seen from these quotations how the limiting experimental difficulties operate to prevent a complete extension of both series even should they be theoretically capable of extension to the fullest extent.

Millikan,[*] in extending the ultra-violet spectrum, remarks that there is evidence for believing that the whole spectrum emitted by the carbon atom up to and including its X-radiation of the L-series has now been obtained. Previously no lines of the L-series of any element of lower atomic number than 30 had been identified.

R. Whiddington [†] points out that since the energy of the $\alpha$-lines in the K and L series is determinable and the speed $v$ of an electron representating the same energy is also determinable, it is possible to formulate a relationship which is fairly well represented by the general formula—

$$v = C(N + D),$$

where N=atomic number, C and D=constants of the K and L series, so that

For the K-series $v = 2(N-2)\,10^8$ cms. per second, and
For the L-series $v = (N-15)\,10^8$ cms. per second.

The characteristic radiations obtained by Moseley, as shown in Fig. 7, are made up of at least five lines in the case of the L-radiation, which lines are designated $\alpha$, $\beta$, $\gamma$, $\delta$ and $\varepsilon$ in the order of decreasing wave-length, as shown by the numbers at the top of Fig. 7.

* *Astrophys. Journ.* (1920), 52, p. 47.      † *Phil. Mag.* (1920), 39, p. 694.

# CHAPTER VI

## A NOTE ON CRYSTAL STRUCTURE

RÖNTGEN-RAYS, or X-rays as they were named by Röntgen himself, being supposed to be of short wave-length, M. v. Laue, in investigating a means of effecting their diffraction in 1912, formulated the method of using these penetrating rays to analyse crystal-lattice structures as first introduced to the scientific world by A. Bravais. This method consisted in directing a small pencil of X-rays through a crystal in a direction parallel to its axis of symmetry, so that the atoms comprising the crystal would act as an *optical grating* (=finely ruled lines on a glass or metal surface) and set up diffraction effects as obtained with ordinary light and a line-ruled grating, and thereby reveal the atoms in their space-lattice arrangement. W. Friedrich and P. Knipping tested this theory in Sommerfeld's laboratory and obtained the results predicted. W. H. and W. L. Bragg have since modified Laue's theory.

FIG. 8.

The action consists in the setting up of secondary wavelets from the atomic centres, which wavelets interfere and give rise to maxima effects. That is to say, when the X-ray pulse strikes into the outer layers of the crystal, considered as being composed of rows or planes rich in atoms, it is reflected in the sense that secondary wavelets are produced which reinforce one another in a selective way (see Huygens' construction described in optical works) and give rise to spots on the photographic plate depicting the positions of the atoms when studied in definite geometrical forms. These 'forms' interpenetrate and overlie one another, but fortunately the method gives a picture of those which fall within the limited selective range of the diffraction and interference phenomena, so to speak. Fig. 8 shows the complex array of spots obtained. These 'Laue spots' appear in practice elliptical, which is due to the angle of the emergent X-ray pencils.

The expression, *a plane rich in atoms*, is best understood by arranging equally-spaced dots on a piece of paper and drawing straight lines through the dots to intersect different numbers.

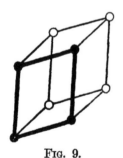

FIG. 9.

The line which cuts through the most dots may be taken as representing the plane richest in atoms. It is owing to such rows that the crystal may be cleaved along their plane. It is evident that the three-dimensional spacing of the atoms introduces certain complications which involve a system of analysis that is not as simple as would be the case if they were studied in one plane.

It is of historical interest to note that Bravais (1848–66) demonstrated that there are only fourteen possible types of symmetrical space-lattices in crystal formation. The *triclinic* type is shown by Fig. 9—after Bravais. Present views differ somewhat from those of Bravais, but this early investigation laid the foundation for the theory of space-lattice structure which X-ray analysis has confirmed.

In Fig. 10 is shown diagrammatically the structure of potassium chloride, as revealed by X-ray analysis, in which the black dots represent the chlorine atoms and the others those of potassium. It is to be noted that the figure only represents

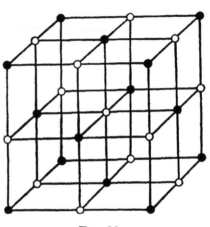

FIG. 10.

an ideal three-dimensional portion removed from a crystal involving many more lattices than shown.

It is to be observed that the molecular individuality of the combination KCl is not realised in the crystal since all the atoms appear to exercise their affinities in all directions—that is to say, in a symmetrical manner. The atoms in this case are probably ions. This feature is discussed in Chapters VII, XIV and XV. See also Appendix VII.

# CHAPTER VII

## RADIO-ACTIVITY AND CRYSTAL STRUCTURE AS CLUES TO ATOMIC STRUCTURE

THE transformation of radium into radium emanation * by the disintegration of the atoms of the former element is a step-process involving the loss of one helium unit of mass 4 from each radium atom, when the action takes place. The respective atomic weights are 226·0 and 222·0. This transformation is, of course, taken as an example. See Appendix VI.

The above change from a soft metal very chemically active, as radium is known to be (analogous to barium), to an inert gas by the loss of one helium unit per atom disintegrating, shows that there must be a change in the disposition of electrons in the atom, as will be seen from the octet theory of atomic structure which is detailed in Chapters XIII, XIV, XV ; but apart from this theory the whole trend of molecular and atomic physics is in the direction that leads to the electrical state of the atom or molecule determining its chemical and physical properties. Moreover, in the case of other radio-active changes, when $\beta$-rays (high-speed electrons : some move with a velocity over $\frac{9}{10}$ths that of light, light velocity being 186,000 miles per second) are ejected from the atom, a transformation of the chemical and physical character of the element also takes place.

This means that a very small change in the disposition of the electrons in the atom completely alters the resultant substance, and therefore it can easily be understood why certain atoms pair and have but little stray field and form in consequence gaseous molecules, whilst others do not pair and the stray fields are such that they agglomerate or cohere in extended mass formation, and assume metallic properties. In the latter case there are electrons in a free state. Crystals are formed which are in a sense mammoth molecules. A crystal of copper, however, is not a non-conductor of electricity like one of sulphur.

Crystal analysis by X-rays shows that the crystal is made up of atoms arranged as points in a space lattice (see Chapter VI), and

---

* Ramsay has suggested the name *niton* (Nt), which is generally accepted, but some experimentalists think the final naming of the radio-atoms should be deferred, so that a proper system of nomenclature can be devised for all members, the present ones being provisional. E. Q. Adams suggests, e.g., " radon," " thoron " and " actinon " for the emanations.

that in certain cases the atoms may be ionised in the polar sense,
much in the same way as they are ionised in an electrolyte, but being
rigidly fixed they cannot give rise to conduction of electricity
by ionic movement; or, considering a substance like sulphur, it
does not afford a mobile electronic state (as is afforded in a copper
crystal, for example) and the substance becomes an insulator. A
more definite statement would be to say that there are two types
of chemical combination distinguishable in crystals, viz. :

(a) Salts, the atoms of which are oppositely charged and
which are held together by the electrostatic force of
attraction existing between the + and − charges.

(b) Atoms which are held together by electrons common to
two nuclei. Diamond and carborundum are of this
type.

The views (a) and (b) have been recently expressed by W. L. Bragg
(1920), Royal Institution Lecture. The octet theory (1916–1919)
embodies similar ideas—see particularly Chapter XV.

Crystal analysis has developed to such an extent that the
reader should consult books devoted exclusively to the subject.
It will serve the present purpose, therefore, to cite some recent
work by W. L. Bragg.*

In studying the arrangement of atoms in crystals, Bragg deduces
the distance between centres of adjacent atoms so that those of a
crystal structure may be pictured as an assemblage of spheres of
diameters characteristic of the atoms, each sphere being, as it were,
held in place by contact with its neighbours. The accompanying
diagram, Fig. 11, shows the values or constants assigned to the
atoms of the elements when arranged in the order of their atomic
numbers, the graph being that of Bragg.

The curve is periodic in character and resembles Lothar Meyer's
curve of atomic volumes. Each atom appears to occupy a constant
space in any crystalline structure of which the atom forms a part.
The space, for example, occupied by the alkali metals (Li, Na, etc.)
is the greatest.

This regularity represents an approximate law with variations
between observed and calculated distances of the order of 10 per
cent. This law is of value in crystal analysis since the conception
of the atoms as a compact assemblage of spheres of known diameters
limits the number of possible arrangements which have to be
tried in interpreting the diffraction of X-rays by the particular
crystal-structure under study.

The significance of the Lewis-Langmuir theory of atomic structure gains appreciation from this investigation; as it follows that
two electronegative atoms are situated close together in crystalline
formation since they share electrons. Consequently, the spheres
should be of small diameter. On the other hand, with electro-

* Phil. Mag. (1920), 40, p. 169.

positive elements (atoms) which do not share electrons in the outer
shell with neighbouring atoms, they should be situated at a greater

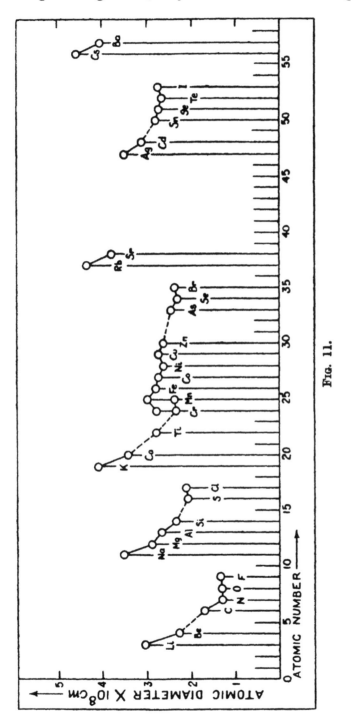

Fig. 11.

distance from one another, and thus they appear to occupy a greater
space in the structure.

Sir W. H. Bragg says : " It is shown that the relation is less accurate when applied to the crystals of metals, which, on Langmuir's theory, consist of an assemblage of positive ions held together by electrons which have no fixed positions in the structure." From the distance between the electronegative atoms holding electrons in common, an estimate is made of the diameter of the outer-electron shells of the inert gases (see Chapters XIV and XV), which in Angström units are : Ne=1·30, Ar=2·05, Kr=2·35, Xe=2·70.

One might here raise the question, as to whether the larger diameter atoms have some nuclear parts or shell-layers in a non-sharing state of ionisation, as distinct from a converse state, for this might possibly account for their greater size if they simulate aggregates with ionised atoms. A further step in this reasoning would be to consider the radio-active atoms as unstable partly on this account (see Chapter XVI).

A. Landé * observes that combinations with the alkali metals and the halogens give the following theoretical results in terms of the grating constants, as determined by X-ray analysis :

TABLE V

| | | | |
|---|---|---|---|
| LiF =4·00 | LiCl =5·11 | LiBr =5·45 | LiI =5·95 |
| NaF=4·60 | NaCl=5·59 | NaBr=5·98 | NaI =6·47 |
| KF =5·31 | KCl =6·24 | KBr =6·59 | KI =7·05 |
| | RbCl=6·57 | RbBr=6·88 | RbI =7·33 |
| | CsCl =6·53 | CsBr =6·81 | CsI =7·23 |

In a further paper in the same journal (2, p. 87) the following values are deduced in which $r$ is the radius of the ion :—

TABLE VI

| Ions= | Na+ | K+ | Rb+ | Cs+ | F− | Cl− | Br− | I− |
|---|---|---|---|---|---|---|---|---|
| $r/0\cdot528 \times 10^{-8}=$ | 2·07 | 2·73 | 3·10 | 3·01 | 2·26 | 3·10 | 3·38 | 3·86 |

It is considered that these values are in accordance with spectroscopic evidence.

The following books may be consulted for further information :—

X-Ray and Crystal Structure (1918), by W. H. and W. L. Bragg.
X-Rays (1918), by G. W. C. Kaye.
Ions, Electrons and Ionising Radiations (1919), by J. A. Crowther.

* Zeits. f. Physik (1920), 1, p. 191.

# CHAPTER VIII

## RUTHERFORD'S NUCLEAR THEORY OF THE ATOM, WITH DETAILED EXPERIMENTAL EVIDENCE: SUPPLEMENTARY NOTES

SIR E. RUTHERFORD on 3 June, 1920, delivered at the Royal Society the Bakerian Lecture,* *On the Nuclear Constitution of Atoms.* A part of the subject treated in this lecture has already been mentioned above, the facts and general theory dating back some years; but additional facts are given in the lecture and much of the existing knowledge is presented in a compact and strongly-convincing form; consequently, it will be desirable to give here the facts and views rather fully, even to the extent of some repetition, since they bear the stamp of probability coming from such a high authority as Rutherford.

By allowing $\alpha$-particles, helium atoms projected with considerable velocity, to strike into matter the remarkable turning-back effect on these particles was revealed. The particles evidently described hyperbolic paths close to the nucleus of the atom encountered, and as the deflection was therefore through a large angle, Rutherford accounted for this circumstance on the assumption that the atom contains a charged massive nucleus of dimensions very small compared with the ordinarily understood magnitude of the atom: or the 'atomic domain,' to use an expression employed by Bragg. It was further assumed that the nucleus had an intense field round it which varied in strength according to an inverse-square law in the region close to the nucleus which rendered the relation between the number of $\alpha$-particles scattered, as found by Geiger and Marsden, consistent with the idea that the resultant positive charge of the nucleus was about $\frac{1}{2}Ae$, where $A$ is the atomic weight and $e$ the fundamental unit charge as given by that of the negative electron. Darwin showed that these results could only follow from the above law of force when taking into account the deflection of the $\alpha$-particles and the mass of the nucleus.

Rutherford's theory involves the idea of the electrically neutral atom having a number of electrons surrounding the nucleus, the number being equal to the number of resultant positive unit charges of the nucleus itself. C. G. Barkla in 1911 had shown from a

* *Roy. Soc. Proc.* (1920), 97, p. 374.

consideration of the scattering of the $\alpha$-rays or particles in experiments of the above kind that the number of external electrons was equal to about half the atomic weight, which was deduced from Thomson's theory that each electron acted as an independent scattering unit. van den Broek had suggested that the nuclear charge was equal to a number assigned to the atoms beginning with hydrogen as 1, helium becoming 2, lithium 3 and so on, up to uranium now known to be 92 ; and Moseley had linked up this number with the simple X-ray spectra law he had obtained from the line-positions of many of the elements, as explained in the previous chapters.

The question whether the atomic number of an element is the actual measure of its nuclear charge is of fundamental importance and all methods of attack should be followed up. Two methods were the scattering of swift $\alpha$- and $\beta$-particles. The former is under further investigation, using the new method, by Chadwick, and the latter by Crowther. The results so far obtained by the first-named experimenter strongly support the numerical equality of the atomic number and the nuclear charge *—within an experimental accuracy of about 1 per cent.

The agreement of experiment with theory for the scattering of $\alpha$-rays, between 5° and 150°, shows that the law of inverse square holds accurately in the case of the heavy elements like gold for distances between about $3 \times 10^{-12}$ cm. and $36 \times 10^{-12}$ cm. from the centre of the nucleus, and, therefore, it may be concluded that there are few if any electrons present in this region. In the case of the lighter atoms, the $\alpha$-particles can approach closer to the nucleus, and in exploring such atoms the field is more promising, as will be seen presently.

The majority of the physical properties of an element appear to be indirectly conditioned by the resultant charge on the nucleus, and consequently the actual mass of the nucleus exercises only a second-order effect on the arrangement of the external electrons and their rates of vibration. The dependence of the properties of an atom on its nuclear charge, and *not on its mass*, then, affords a rational explanation of the existence of isotopes.

It is of interest to note the very different rôle played by the electrons in the outer and inner parts of the atom, as in the former the electrons arrange themselves at a distance from the nucleus, controlled no doubt mainly by the resultant charge of the nucleus and the interaction of their own electric fields. In the case of the nuclear electrons they form a very close and powerful combination with the positively charged units that go to make up the main mass of the nucleus, and as far as is known there is a region just outside the nucleus where no electron is in stable equilibrium. " While no doubt each of the external electrons acts as a point charge in considering the forces between it and the nucleus, this

* See Chadwick's experiment, Chapter III.

cannot be the case for the electron in the nucleus itself. . . . Under the intense forces in the latter the electrons are much deformed and the forces may be of a very different character from those to be expected from an undeformed electron as in the outer atom. It may be for this reason the electron can play such a different part in the two cases and yet form stable systems." The fields due to electrons in motion have to be taken into account.

The hydrogen nucleus is expected to be of the simplest type, and if it be the positive electron its dimensions may be exceedingly small compared with the negative electron (see Chapter III). Helium atoms, or $\alpha$-particles, have a more complex nucleus, and the space occupied by the nucleus is no doubt greater as it becomes more complex. The diameter of the nuclei of light atoms except hydrogen is probably of the order of $5 \times 10^{-13}$ cm., and when two such nuclei collide they may, should the force be sufficient, penetrate each other's structure. Under such conditions only very stable nuclei would be expected to survive the collision undamaged. It is thus of great interest to examine whether evidence can be obtained of their disintegration.

Evidence has already been given * that the passage of swift $\alpha$-particles through dry nitrogen gives rise to fast-moving particles that closely resembled hydrogen in the brilliancy of their scintillations when allowed to strike against a zinc-sulphide screen, and the range of these particles corresponded also with that of hydrogen. The general evidence was that the particles were hydrogen atoms which had their origin in the nitrogen atoms. New apparatus is described for deflecting the particles by a magnetic field. Radium C was used, as before, for the source of high-speed $\alpha$-particles. The apparatus is simple in construction, consisting of a rectangular brass box with a window at one end where a zinc-sulphide screen was placed. The radium C was lodged on one end of a plate (farthest end from the screen) forming a table parallel with the floor of the box. Another parallel plate was placed 8 mm. distant, the two together forming a sort of miniature dinner-wagon, these plates being 6 cms. long by 1·5 cms. wide. The box was placed between the poles of a large electromagnet so that the magnetic lines of force would be in a direction to deflect those particles projected towards the screen either downwards or upwards according to the direction of the field. One edge of the plates almost touched the screen. When the apparatus was in operation the air was nearly exhausted therefrom, leaving a pressure of a few centimetres. When the field was off, the edge of the bottom plate nearest to the screen formed a shadow-line or edge on the screen by cutting off sharply the particle-rays, these being made visible by the scintillations of the zinc sulphide. A microscope was provided and so adjusted that its horizontal hair (usually referred to as

* Rutherford, *Phil. Mag.* (1919), 37, p. 537, Parts I, II, III and IV ; see particularly Part IV

' cross-hairs ') coincided with the ' shadow-edge.' On exciting the magnet, when the rays were bent upwards, scintillations appeared below the hair-line. On reversing the lines of the magnetic field (calling this the minus field) the rays were bent downwards and scintillations appeared on the screen above the hair. The strength of the field was, however, adjusted so that with a minus field the scintillations were observed all over the screen, while with the plus field they were mainly confined to a region below the hair-line. The appearance in the field of view is shown by Figs. 12 and 13, the dots representing the scintillations.

" Since the number of scintillations in the experiments with nitrogen was much too small to mark directly the boundary of the scintillations, in order to estimate the bending of the rays, it was necessary to determine the ratio of the number of scintillations with the + and − field."

" The position of the microscope and the strength of the magnetic field were in most experiments so adjusted that this ratio was about one-third. Preliminary observations showed that this ratio

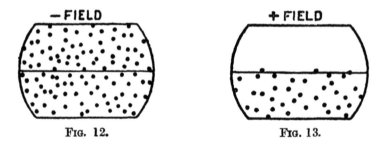

Fig. 12.        Fig. 13.

was sensitive to changes of the field, and it thus afforded a suitable method for estimating the relative bending of any radiations under examination."

After the position of the microscope was fixed, air was admitted to the box, and a continuous flow of dry air maintained through it. An absorbing screen was introduced at the edge of the plates next to the screen to stop atoms of nitrogen or oxygen of range 9 cms. The number of scintillations was then counted for the two directions of the field. The deflection due to the unknown radiation was directly compared with that produced by a known radiation of α-rays from thorium C of range 8·6 cms. The numbers of scintillations with the + and − field were determined as before.

Passing over some experimental values which led to the conclusion that the projected particles could not be helium atoms carrying *one* positive charge, a test was made by introducing one volume of hydrogen to two volumes of carbon dioxide into the box, and after circulating this mixture through it the proportions of the two gases were so adjusted that the stopping power of the mixture for α-rays was equal to that of air. Under these conditions, the H-atoms, like the nitrogen particles (those from nitrogen), are produced throughout the volume of the gas, and probably the relative distribution of H-atoms along the path of the $\alpha$-rays is not very different from that of the nitrogen particles when under examination. If the nitrogen particles are H-atoms, the average deflection should be nearly the same for the H-atoms liberated from the hydrogen molecules ($H_2$) of the mixture, as with the nitrogen. Experiments showed that the ratio of the number of scintillations in the + and − fields of equal value was so nearly identical in the two cases that it was impossible to distinguish between them. The conditions being therefore the same the equality in the ratios showed *that the long-range particles liberated or disrupted from nitrogen were atoms of hydrogen, and the possibility of these long-range particles being of mass 2, 3 or 4 was definitely excluded.* One must pause here to realise what this means : *It is the partial breaking up of the nitrogen atom—one of the great achievements of science.*

Thus it will be seen that the previous 1919 experiment * is in general confirmed, and for the first time the disintegration of the nitrogen atom by swift $\alpha$-particles is placed on record as an accomplished fact. While the effect is small, only one $\alpha$-particle in 300,000 gets near enough to the nucleus of the nitrogen atom to displace a hydrogen sub-atom, it is of far-reaching theoretical consequence.

Sir E. Rutherford says : " It may be possible that the collision of an $\alpha$-particle is effective in liberating hydrogen from the nucleus without necessarily giving it sufficient velocity to be detected by scintillations. If this should prove the case, the amount of disintegration may be much greater than the value given above."

In pursuing these researches further, experiments were made with solid bodies containing nitrogen in combination : boron nitride, sodium nitride, titarium nitride and para-cyanogen. The method of analysis was much the same as described above, and in all the experiments the masses liberated corresponded with the H-atom, no particles of mass 2, 3 or 4 being detected. As a control experiment, substances were tried which did not contain nitrogen, and no H-atoms were detached under the conditions of the experiment. In the case of boron nitride there is, however, the uncertainty that the boron itself may emit H-atoms on bombardment with $\alpha$-particles, and on this account and others the experiments on solid nitrogen compounds were abandoned for the present.

In the above experiments with gaseous nitrogen, the H-atoms driven from the N-atom were of long range. A shorter-range effect was noticed, but this was previously attributed to oxygen and nitrogen set in swift motion with a single charge on each atom in close collision with $\alpha$-particles. With the wide-slit † arrangement

---

* See previous citation : *Phil. Mag.* (1919).

† The gap between the two plates (or tables of the dinner-wagon) is called the slit.

as described above, it was found that the short-range effect was due to a mass lying between 1 and 4, and from the range of the particles and the amount of their deflection that the atom, moreover, carried *two* positive units of charge.

In a more decisive test of the deflection of the oxygen atoms in $+$ and $-$ magnetic fields, a mixture of hydrogen and carbonic acid gas was used for comparison. From these experiments and the range of the oxygen atoms in air the mass of the particles, in this case disrupted from the oxygen atom, was obtained.

It was found that the path of the atoms from oxygen was bent about 5 per cent less than that of the H-atoms. Also from a previous paper,* reasons were deduced for believing that the range $x$ of the particles from oxygen is given by the equation—

$$\frac{x}{R} = \frac{m}{M}\left(\frac{u}{V}\right)^3.$$

The $\alpha$-particle data being—

R = range.
M = mass.
V = velocity.

The oxygen-particle data being—

$x$ = range.
$m$ = mass.
$u$ = velocity.

Since $x = 9 \cdot 0$ cms. range for the particles or atoms from oxygen disrupted or disintegrated by the colliding $\alpha$-particles from radium C of range 7 cms.

$$\frac{x}{R} = 1 \cdot 29,$$

and taking M $= 4$

$$mu^3 = 5 \cdot 16 v^3.$$

For further particulars see original papers.† Suffice it to say, this type of calculation has been shown to account for the range of the H-atoms, and it is believed that it would be fairly accurate for those of shorter range which are not very different. Having found that $1 \cdot 14 MV = 1 \cdot 25 mu$, and that $mu^3 = 5 \cdot 16 v^3$,

$$u = 1 \cdot 19V \text{ and}$$
$$m = 3 \cdot 1.$$

"Considering the difficulty of obtaining accurate data, the value $m = 3 \cdot 1$ indicates that the atom has a mass of about 3, and this value will be taken as the probable value in later discussions."

When air was substituted for oxygen the bending of the path

* Previous citation : *Phil. Mag.* (1919).
† *Roy. Soc. Proc.* (1920), **97**, p. 390; and *Phil. Mag.* (1919), **37**, p. 571.

of the particles, i.e. the curved deflection, was the same, indicating that the short-range atoms from oxygen and nitrogen have a mass of 3 and carry in each case a double positive charge, and they are projected with a velocity 1·19V, where V is the velocity of the colliding $\alpha$-particle of mass 4. Since the masses 3 and 4 carry a double charge Rutherford regards them as being isotopic, so that in a nuclear structure these masses might perform like functions (see below).

It is thus evident that in the nuclei of oxygen and nitrogen atoms there are sub-atomic masses 3; moreover, in the case of nitrogen the sub-atom of mass 1 is also contained. The term *sub-atom* is here used to distinguish these parts evolved from normal atoms, as the former becomes an atomic part when considered bound together with electrons or positive and negative charges.

It is clear from the above that the nitrogen nucleus can be disintegrated in two ways, viz.—

> (i) By the expulsion of a mass of 1.
> (ii) By the expulsion of a mass of 3.

It is considered that these two forms of disintegration are independent of each other and that they are not simultaneous effects.

Under the title ' energy considerations,' some very interesting deductions are made. It is hoped to get further information as to the energy conditions which determine the disintegration of the atoms by studying the ionisation tracks by the method of C. T. R. Wilson (see Chapter III).

Experiments are being undertaken to ascertain whether $\beta$- and $\gamma$-rays, and cathode rays, will effect the disintegration of the atoms as in the above experiments with $\alpha$-particles—at least certain of these sources of power are to be tried.

Discussing the foregoing results, it is shown that the atoms of nitrogen and oxygen contain sub-atoms or nuclei of mass 3 and, in the case of nitrogen, two sub-atoms are contained, these being masses 1 and 3. Since the mass 3 has two positive charges, the same as helium nucleus of mass 4, it may be assumed that they are chemically alike, in fact the two masses may be isotopic. It would, moreover, be expected that their spectra should be nearly alike, and, indeed, *all* their properties both physical and chemical should nearly coincide.

In considering the possible constitution of the atoms of the elements, it is natural to suppose that they are built up ultimately of hydrogen nuclei and electrons. On this view the helium nucleus is composed of four hydrogen nuclei and two negative electrons with a resultant charge of two positive units of electricity.

One is reminded here of Prout's hypothesis, advanced over one hundred years ago, that all the elements were built up of hydrogen as a primal unit. The work of Aston (see Chapter II) in showing that all the atoms of all the elements (except hydrogen) are whole numbers gives support to this type of theory, and the experiments recorded above without exception give strong evidence as to the unity idea of matter. Prout did not know the part electricity would play in the atom, consequently his theory is more of the nature of an idea, if one can so express it. Sir Wm. Crookes, many years ago, made a very prophetic statement to the effect that the atoms were of electrical origin.* Here was another ' idea ' which has proved to be true. It must be remembered that the state of knowledge was not such at the times of Prout and Crookes to enable them to develop their ideas into what might be termed a proper theory. A cautious observer might justly remark that these earlier ideas are close enough to the truth to be coupled up as forerunners to present-day knowledge of the atom ; but he might observe that it is hardly safe to assume that all the atoms are polymerides of hydrogen, yet the nuclear theory of matter goes so far in that direction that one may accept Prout's hypothesis as being at least probably true in the main (see Chapter XX).

Returning to the lecture, the mass of helium being 4·00, in terms of oxygen 16, it is thus less than four hydrogen atoms, viz. 4·032, the mass of hydrogen being 1·008 ; but this discrepancy is generally supposed to be due to the close interaction of the electric fields in the nucleus, resulting in a smaller electromagnetic mass than the sum of the individual components. A. Sommerfeld † has concluded from this consideration that the helium atom must represent a very stable structure, and this agrees with experiment, as no case is on record of this atom being broken up by any process of dis-integration.

Aston has shown (Chapter II) that all the atomic weights thus far measured are whole numbers within the limits of experi-mental error, with, of course, the exception of hydrogen, which is 1·008.‡

Rutherford says : " If we are correct in this assumption," referring to the possibility that the elements are mainly built up of helium and its isotope as secondary units, the primary ones being hydrogen groupings, " it seems very likely that one electron can bind two H-nuclei and possibly one H-nucleus. In one case this entails the possible existence of an atom of mass nearly 2 " (it might be a shade over this value, for example) " carrying one charge, which is to be regarded as an isotope of hydrogen. In the other case, it involves the idea of the possible existence of an atom of mass 1 which has zero nuclear charge. Such an atomic structure seems by no means impossible.

---

* See *Chem. News*, vol. 55, pp. 83 and 95.

† *Atombau und Spektrallinien* (1919), p. 538. See 2nd edition (1921), p. 569 ; see page 153 of this book.

‡ Aston's $H_3$, by the way, has a mass of $3 \times 1·008$. J. J. Thomson discovered this molecular analogue of ozone (see p. 206).

On present views, the neutral hydrogen atom is regarded as a nucleus of unit charge with an electron attached at a distance, and the spectrum of hydrogen is ascribed to the movements of this distant-electron. Under some conditions, however, it may be possible for an electron to combine much more closely with the H-nucleus, forming a kind of neutral doublet. Such an atom would have novel properties. Its external field would be practically zero, except very close up to the nucleus, and in consequence it should be able to move freely through matter. Its presence would probably be difficult to detect by the spectroscope, and it may be impossible to contain it in a sealed vessel. On the other hand, it should enter readily into the structure of atoms, and may either unite with the nucleus or be disintegrated by its intense field, resulting possibly in the escape of a charged H-atom or an electron or both." See in this connection Chapter XX.

It seems safe to conclude that atomic nuclei may be made up of masses

1, 2, 3 and 4

and even greater masses functioning as secondary units. At any rate the nuclei thus far recognised experimentally may be represented thus—

$$\overset{+}{\text{H}} \text{ mass 1,} \quad \overset{++}{\text{X}} \text{ mass 3,} \quad \overset{++}{\text{He}} \text{ mass 4.}$$
(See Supplementary Notes below.)

It is possible now to build up some of the lighter atoms from these nuclear units and electrons, bearing in mind that when the resultant charge is the same on different atoms so constructed they are isotopic.

The procedure is of course tentative, and Rutherford recognises that it would be premature to press these ideas too far in those cases pending the discovery of isotopes. The following examples, however, should be of suggestive interest.

The element lithium may consist of a mixture of isotopic atoms made up of units, thus :—

$$\left.\begin{array}{cc} + & + \\ 3 - & 3 \\ + & + \end{array}\right\} = \text{mass 6}$$

$$\left.\begin{array}{cc} + & + \\ 3 - & 4 \\ + & + \end{array}\right\} = \text{mass 7}$$

$$\left.\begin{array}{cc} + & + \\ 4 - & 4 \\ + & + \end{array}\right\} = \text{mass 8}$$

A mixture of these isotopic atoms giving a mean value of 6·96, with a nuclear charge of 3 on each atom, which is the atomic number, *represents the element lithium.*

In the above scheme and in those following the signs and figures have the following significance :—

$$- = 1 \text{ binding electron.}$$
$$- \; - = 2 \text{ binding electrons.}$$
$$- \; - \; - = 3 \text{ binding electrons.}$$
$$+ = 1 \text{ positive charge of the nuclei.}$$
$$+ \; + = 2 \text{ positive charges of the nuclei.}$$
$$1, 3 \text{ and } 4 = \text{masses of respective nuclei (atoms of sub-atomic character).}$$

The nucleus of the carbon atom would be represented thus :—

$$\left.\begin{array}{cc} \overset{++}{3} & \overset{++}{3} \\ \quad -\;- \\ 3 & 3 \\ ++ & ++ \end{array}\right\} = \begin{array}{l} \text{mass 12 and charge 6, which is the} \\ \text{atomic number.} \end{array}$$

In this case, as with those following, there are no isotopes (see Table II, p. 14), consequently all the atoms are alike in mass.

The element nitrogen is of interest as it would have, according to the above-described experiments, masses of 1 carrying a single + charge each, whilst the other parts would be as shown above.

$$\left.\begin{array}{cc} \overset{++}{3} & \overset{++}{3} \\ + \quad + \\ -\; 1-\; 1- \\ 3 \quad\quad 3 \\ ++ \quad ++ \end{array}\right\} = \begin{array}{l} \text{nitrogen atom of mass 14, charge 7,} \\ \text{which is the atomic number.} \end{array}$$

It will be seen that the particles of mass 3 (nuclei) are introduced in this structure, and these are indicated by the experiments detailed above.

In a similar manner the oxygen atom may be represented thus :—

$$\left.\begin{array}{cc} \overset{++}{3} & \overset{++}{3} \\ ++ \\ -\quad 4 \quad - \\ 3 \quad\quad 3 \\ ++ \quad ++ \end{array}\right\} = \begin{array}{l} \text{oxygen atom of mass 16, charge 8, which} \\ \text{is the atomic number.} \end{array}$$

Referring to these schemes Rutherford says : " The carbon nucleus is taken to consist of four atoms of mass 3 and charge 2, and two binding electrons. The change to nitrogen is represented by the addition of two H-atoms with a binding electron." The oxygen nucleus is obtained by the substitution of a helium nucleus in place of the two H-atoms (see Chapter XVI).

From the structure of nitrogen it will be seen that the masses of 3 are more readily hit as shown in the bombardment experiments above, than the more secluded H-atoms, and this would seem to suggest why more of the former are liberated. The actual arrangements of the nuclear atoms and electrons in the above structures are somewhat diagrammatic as the configurations are not known.

Further experiments are contemplated and it is intended to extend the treatment to other light atoms ; but unless they are gaseous it is difficult—

(i) To ensure absence of hydrogen as an impurity, and
(ii) To prepare sufficiently uniform films when dealing with solids.

### SUPPLEMENTARY NOTES

1. At the meeting of the British Association at Cardiff, it was announced by Rutherford that hydrogen of mass 1 was probably a disintegration product of nitrogen by the method given above, whilst from oxygen ($O_2$) and carbon dioxide ($CO_2$) particles of masses 3 and 2 were driven from the atoms, but no particles were observed having a mass of 1 in this case, i.e. when bombarding $O_2$- and $CO_2$-molecules. When the bombardment takes place it is against the atom of the molecule—that is to say, when a hit is scored. *The liberation of sub-atoms of mass 2 constitutes a new fact in addition to those given above.*

*The sub-atomic or nuclear parts of atoms thus far discovered experimentally have the masses* 1, 2, 3 *and* 4, *the last-named one occurring in radio-active changes.*

2. In a later communication by Rutherford and Chadwick * the following experimental results are recorded : It had been found that the long-range particles from nitrogen were charged hydrogen atoms which indicated that some of the nitrogen atoms were disintegrated by the $\alpha$-particles. Recent improvements in the optical conditions have been made, whereby the counting of the scintillations on the zinc-sulphide screen has been made easier, and the results now obtained are considered more reliable. They had been able to show definitely that the H-atoms from nitrogen have a greater range than those from hydrogen. The H-atoms driven from the hydrogen molecule or from a hydrogen compound have a maximum range of 29 cms. of air, while the H-atoms from nitrogen have a greater range of 40 cms., giving a ratio of 1 to 1·4. The range of the bombarding $\alpha$-particles in these experiments was 7 cms. " This result shows that these particles cannot possibly arise from any hydrogen contamination."

* *Nature* (1921), 107, p. 41.

Other elements were subjected to the $\alpha$-particle bombardment, with the result that *long-range* H-atoms were liberated from

> Boron,                     Sodium,
> Fluorine,                  Aluminium and
>              Phosphorus ;

but the number from boron and sodium was much fewer than from the other three elements. It had been found previously that the *long-range* H-particles were not liberated from either oxygen or carbon-dioxide molecules.

With solids, the bombardment was against a thin film of the element or its oxide. Observations of the number of scintillations were made through a thickness of mica corresponding to a distance of 32 cms. of air, and the results are said to be independent of the presence of hydrogen or any hydrogen compound in the material.

The following elements were bombarded, and they showed very little, if any, effect with an absorption corresponding to 32 cms. of air : Li, Be, C, O, Mg, Si, S, Cl, K, Ca, Ti, Mn, Fe, Cu, Sn and Au.

The gases $O_2$, $CO_2$ and $SO_2$ were examined for H-particles. At absorptions *less* than 32 cms. of air no traces of H-atoms were obtained. Elements other than the foregoing have not yet been examined for particles having a range less than 32 cms. in air.

The particles from all the first-mentioned elements have a maximum range of at least 40 cms. The particles from aluminium have a surprisingly long range of 80 cms.

While there is no experimental evidence of the nature of these particles, except in the case of nitrogen, it seems probable that they are all H-atoms liberated at different speeds. Assuming the law connecting the range and velocity of the H-particles to be the same as that for the $\alpha$-particles, it would appear that the energy of the H-particles from aluminium is about 25 per cent greater than the energy of the incident $\alpha$-particle.

Rutherford and Chadwick remark : " It is of interest to note that no effect is observed in ' pure ' elements [those without isotopes such as C, N, O and F] the atomic mass of which is given by $4n$, where $n$ is a whole number. The effect is, however, marked in many of the elements the mass of which is given by $4n+2$, or $4n+3$. Such a result is to be anticipated if atoms of the $4n$ type are built up of stable helium nuclei and those of the $4n+a$ type of helium and hydrogen nuclei.

" It should also be mentioned that no particles have so far been observed for any element of mass greater than 31. If this proves to be general, even for $\alpha$-particles of greater velocity than those of radium C, it may be an indication that the structure of the atomic nucleus undergoes some marked change at this point ; for example, in the lighter atoms the hydrogen nuclei may be satellites of the main body of the nucleus, while in the heavier elements the hydrogen nuclei may form part of the interior structure."

This experimental work is of very far-reaching significance, and it is especially important in affording a powerful tool for investigating sub-atomic phenomena and revealing secrets which some years ago one would have considered beyond the power of man to discover. The long-range particles from aluminium seem to show that we are on the track of setting free atomic energy on a larger scale than hitherto found possible (see Chapter XXI) ; but as yet negligibly small from a practical point of view.

3. Now that Aston * has been able to isolate by the positive-ray method some of the isotopes of the alkali metals by using a heated anode as a source of the corpuscular rays,† we seem to be on the eve of clearing up at least one or two outstanding problems of chemistry, the cause of fractional atomic weight values in particular ; and at the same time more definite knowledge of the atom itself is revealed. It would appear that lithium has isotopes of masses 6 and 7, potassium 39 and 41, rubidium 85 and 87. Sodium gives a single line answering to the atomic mass of 23. If no other mass is discovered, sodium would then be known as a ' pure ' element. The values for the lithium atoms are of interest in connection with one of Rutherford's nuclear schemes given above.

4. The term *proton* has been adopted by some authorities as a name for the positive electron, or the positive entity, as, for example, the hydrogen atom without its normally attendant electron. Atoms are supposed to be made up of negative electrons and protons, as already indicated.

Rutherford, speaking cautiously, remarks on the probability of the proton not being a true counterpart of the negative electron, since the former seems in particular to involve a relatively large mass compared with the mass of the negative electron. Perhaps it is in the nature of things that there should not be a true counterpart ; and there may thus be a difference in more ways than one.

---

* *Nature* (17 March, 1921).

† See O. W. Richardson, *Emission of Electricity from Hot Bodies* ; see also G. P. Thomson, *Camb. Phil. Soc. Proc.* (1920), 20, p. 210. Gehrcke and Reichenheim (*Ver. d. Phys. Gesell.* (1906), 8, p. 559 ; (1907), 9, pp. 76, 200, 376 ; (1908), 10, p. 217) first obtained positive rays with the alkali metals.

# CHAPTER IX

## THE QUANTUM THEORY

In studying the interchanges of radiant energy between material bodies and the surrounding space containing radiant energy, it was found that, in order to have a state of equilibrium in which as much energy was absorbed by a given body as re-emitted by it as radiation, the radiant energy in equilibrium must be considered as finite and not infinite. If infinite, there would be a perpetual loss of radiation, as in infinite space, whereas in suitably-constructed enclosures this is not the case. This leads to the conclusion that Newton's laws fail in extreme cases, and it is worthy of note that in this circumstance the magnitudes taken individually are physically very small.

According to Newtonian dynamics the energy would degrade into the shortest vibrations the material medium is capable of sustaining ; whereas with radiation there is no quickening of the vibrations in the medium exterior to the material body at the expense of the energy of the body, though the external medium may be capable of executing the shortest possible vibrations. In short, the Newtonian laws do not lead to a solution of such thermodynamic phenomena.

The radiation emitted by a black body (=a full radiator) can be analysed by means of a prism, in which case the spectrum produced is a continuous band. The spectrum is not, of course, in the visible region when the body is *black* in the ordinary sense of the term, but at high temperatures it emits visible radiations. By studying the energy of this radiation or spectrum, it was found that the density of the radiation (quantity of energy in a cubic centimetre) emitted by oscillators or vibrators (these may be taken as atoms) was proportional to the mean energy of the oscillators, these being in statistical equilibrium with the radiation ; and as the radiation density becomes infinitely small for short wave-lengths, the mean energy also becomes exceedingly small. When the frequency is very great the energy of the oscillators may thus be very small—the radiation density being still proportional to the energy.

Now considering such radiation in thermal equilibrium with gas molecules, it will be independent of the period involved in the kinetic energy of these molecules, as such activity would sustain considerable oscillatory energy, and the impossible state would be arrived at whereby the radiation is infinite for small wave-lengths. Thus taking a small amount of radiant energy, to maintain thermal-equilibrium, when for example very great kinetic activity, but low temperature, is involved in the gaseous medium exterior to the oscillators of the solid, the energy cannot be shared, as it were, with the whole number of molecules so that it will be everywhere the same at every instant of time. The *measurable* temperature, whatever that may be, is the same throughout the enclosure. Consequently, the energy has to be ' bunched up,' or emitted spasmodically, to maintain the equilibrium principle, and this leads to the *discontinuity* involved in the quantum theory.

It is necessary therefore to assume that the energy of the oscillators varies in a discontinuous manner, and Planck supposes that it varies by equal quanta represented by $h\nu$, $\nu$ being the number of vibrations per second and $h$ the universal constant that bears Planck's name. The energy, or strictly speaking the ' action,' thus becomes *atomic*. Whether the atomicity is due to the oscillators parcelling out energy in quanta, or whether it is due to an inherent property of energy itself, is not perhaps known with absolute certainty ; but there are indications that the energy or action is really atomic, although Planck in a further development of his theory supposes that the radiant energy is absorbed continuously while it is emitted in jumps.

Thus, when radiation is given out, it occurs in quanta and each amount forms what might be termed *a bundle of energy*—to use a well-known expression.

The magnitude of this quantum or bundle varies continuously with the frequency of the radiation by definite increments, so that the energy E of a quantum for a given frequency is

$$\text{E} * = h\nu$$

The value $h$ has been determined by radiation formulæ and it is always

$$6{\cdot}55 \times 10^{-27}.$$

This quantum of energy has been found to occur in the formulisation of such phenomena as—

> Black-body radiation.
> Specific heats of solids.
> Photo-electric effect.
> Line-spectra of hydrogen and helium (see Chapter X).

In the introductory statements above no special reference was made to the *æther*, but the term *medium* may be taken to represent

---

* The value of E varies from $1{\cdot}9 \times 10^{-11}$ to $2{\cdot}1 \times 10^{-14}$ erg, according to the vibration frequency of the resonator or oscillator. This vibration frequency is of course that of the radiation as well.

it where the context admits. The æther has to be considered, and it may be conducive to clearness to restate a portion of the above so as to include specially the æther.

If very large or massive molecules are mixed with exceedingly small or light ones, it becomes evident from the Newtonian laws, as applied to gases, that the kinetic energy of the system as a whole resides largely in the small molecules if they are more numerous ; the *degrees of freedom* being the same in both cases.

It is assumed that the small molecules approach more nearly to the character of the æther than the large ones, and by considering a very great number of such small molecules, so as to present a continuous medium, the condition of the æther may be approximated.

It is, moreover, assumed with equal probability that the æther would not have fewer degrees of freedom than the molecules, but in fact more degrees as a result of its continuity or fine-grained structure. It would not perhaps be pushing matters too far to assign to it an infinite number of degrees of freedom.

This being the case, and assuming the *equipartition of energy* * as involved in mechanics, practically all the energy should be taken up by the æther *which is not in accordance with observation.*

The foregoing statements may be supported by experiment for, in the case of a suitable enclosure with a central body (corresponding to the few massive molecules above) raised to a given temperature, there is evidence that the medium in the space (representing the æther or the small molecules) between the walls of the enclosure and the central body is in temperature-equilibrium with its surroundings, though it should *not* be if the Newtonian laws apply.

Lord Rayleigh and J. H. Jeans have shown that the law of equipartition of energy involving the æther in the case of the black-body spectrum (emission) leads to a formula which should apply in this circumstance, but it does not agree with experiment. Planck introduced the idea of discontinuity somewhere in the emission or absorption of energy in the system (as already stated), which overcomes the difficulty.

The writer has drawn somewhat freely from Chapter VI of Perrin's book *Atoms* † in preparing some portions of the above earlier parts of this chapter. The following taken from Jeans' *Report on Radiation and the Quantum Theory* will give the reader a further insight into this problem : " . . . in all known media there

---

* The principle of equipartition of energy, as the name implies, involves an exchange of energy so that its distribution will be equally apportioned (statistical *means* involved in the calculations) amongst the systems in movement whereby their degrees of freedom will afford an equal sharing of the activity after a steady state has been reached. This idea is usually expressed by the statement that " the energy is equally divided among all the degrees of freedom." This principle applies to all bodies capable of vibration or movement, but it may not apply to negative electrons. Whether it applies to the æther is not a settled question.

† Translation by D. Ll. Hammick.

is a tendency for the energy of any systems moving in the medium to be transferred to the medium and ultimately to be found, when a steady state has been reached, in the shortest vibrations of which the medium is capable. This tendency can be shown to be a direct consequence of the Newtonian laws. This tendency is not observed in the crucial phenomenon of radiation ; the inference is that the radiation phenomenon is determined by laws other than the Newtonian laws. . . . The tendency of the energy to run into the vibrations of the shortest wave-length is found to admit of rigorous proof. The second step is really the essential one. It consists in working back from the observed final partition of energy in the æther to the laws by which the partition of energy must be produced. It can be shown that, if the final partition of energy is that given by the well-known law of Planck, then the motion of the medium must be governed by laws which involve the quantum theory " (p. 7).

See also *A System of Physical Chemistry*, vol. iii. (1919), by W. C. M·C. Lewis.

# CHAPTER X

## THE BOHR-RUTHERFORD ATOM: SOMMERFELD'S EXTENSION OF THIS THEORY: AND EPSTEIN'S STUDIES OF THE STARK EFFECT

PROF. SIR E. RUTHERFORD'S views of the nuclear constitution of the atom, as given in Chapter VIII, had been developing for some years; arising, it would seem, out of the ideas of Lord Kelvin, H. Nagaoka, Sir J. J. Thomson, and the phenomena of radio-activity; but in particular based upon the experiments of Geiger and Marsden, as already stated.

Dr. Neils Bohr, of Copenhagen, took the matter up from the point of view of the optical spectra, taking hydrogen and helium as affording the simplest types of atom to study.

In the laboratory the spectroscope ordinarily reveals in the case of hydrogen 12 lines; in the stars there appear to be 33 lines; whilst in a nebular photograph, taken by W. H. Wright at the Lick Observatory, 11 lines show clearly on the plate. R. W. Wood [*] has recently succeeded in extending the hydrogen lines, that is to say the Balmer series, with a new type of discharge tube whereby the 20th line has now been recorded. Bohr had assumed that in order to have a larger number of lines than hitherto found in the laboratory experiments the atoms must be far enough apart to allow for the existence of large orbits, as will be seen from the following.

In order to account for the hydrogen lines (Balmer series) and their positions, Bohr assumed that the electron of the hydrogen atom was in rapid revolution in a circular orbit round the positive nucleus of the atom, and that the monochromatic radiations—those which give the lines, each line being the image of the slit of the spectroscope—occurred only when the electron made an excursion from an outer orbit to an inner one, the orbits being ideally represented by Fig. 14 (see p. 96 for spacing of orbits).

It was during this transition state that the electron was supposed to give out the radiation. So long as the electron continued to revolve in a given orbit under uniform angular velocity no radiation

[*] *Roy. Soc. Proc.* (1920), **97**, p. 455.

was supposed to take place. Now these stable states were arrived at on the assumption that the centrifugal force due to the speed of revolution of the electron exactly balanced the electrostatic attraction of the positive nucleus—that is to say, the mutual attraction between the negative electron and the nucleus. A further circumstance, however, had to be taken into account, as from the behaviour of charges of electricity, according to the electromagnetic theory, radiation should take place, even should the speed of revolution be uniform, and as a consequence of this loss of energy the electron should fall into the nucleus. To get over this difficulty of the non-radiant state of the atom, but involving a uniformly revolving electron, it was assumed that radiation could only occur in jumps or quanta according to Planck's theory of radiant-energy exchanges or emissions, thus securing a necessary state of stability until a definite emission took place. This difficulty may, of course, have been an imaginary one, since the mechanism of radiant energy is not known, and it is even pos-sible to have electrons revolv-ing between atomic nuclei, as suggested by Bohr in the case of the hydrogen molecule, and re-volving in such a way that their lines of force will join opposite nuclei while the plane of re-volution might be about mid-way between them. Assuming that there are no other lines of force except those terminating in opposite nuclei, there would be no occasion to consider a light undulation being set up

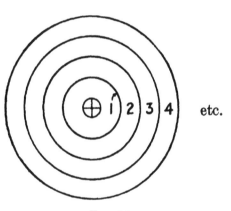

FIG. 14.

with such a system ; but on removing one of the nuclei the exposed lines of force could then extend into space or couple themselves up with a space-field already in existence, and give rise to a disturbance known as an electromagnetic wave. This may be otherwise described as a process of ionisation (see Chapter XII). From the point of view involving the angular-momentum idea, which is the nature of energy multiplied by time (see below), the contraction of the orbit would perhaps come to the same thing, this contraction taking place when a radiation disturbance of electromagnetic origin occurs. Light is, of course, known to be an electromagnetic wave pheno-menon.*

However this may be, returning to the Bohr theory, the angular momentum of the revolving electron is its velocity multiplied by its mass taken in a curved path—that is to say, through an angle ; but since the radius of the circular path is important it is really

* See Preston's *Theory of Light.*

the product of this radius ($r$) and its mass ($m$) multiplied by its velocity—this velocity equalling

$$\frac{1}{r}\times\frac{h}{2\pi m}$$

Angular momentum is therefore a term that summarises the energy conditions entering into the moving electron, and taking this energy as momentum units helps to get over the difficulty mentioned above.

Each one of the states 1, 2, 3, 4, etc., in Fig. 14 answers to one in which the angular momentum is an exact multiple of Planck's constant $h$ (the value of $h$ being $6\cdot55\times10^{-27}$ erg second), that is to say, it must be, for example, $1\times h$ or $2\times h$ or $3\times h$, etc.—but higher multiplying numbers ($n$) occur in practice—so that

$$2\pi h \times \text{angular momentum}=nh,$$

the angular momentum of each electron *during its steady state* being a whole-number multiple of

$$h/2\pi$$

$\pi$ is, of course, the ratio of the diameter of a circle to its circumference $=3\cdot14159$ . . .

The expression answering to angular momentum leads towards a physical interpretation of Planck's constant $h$. For, if the angular momentum is represented by M, it follows that for a circular orbit

$$\pi M=\frac{T}{w},$$

where $w$ is the frequency of revolution of the electron in its circular orbit and T its kinetic energy. Now

$$h/2\pi=1\cdot04\times10^{-27}$$

which may be taken as a sort of unit, or single quantum of angular momentum, so that

$$M=n\times1\cdot04\times10^{-27} \; exactly$$

for each steady state *independent of the charge on the nucleus*. This is perhaps tantamount to saying that " When a body rotates at the rate of $w$ revolutions per second, its energy is equivalent to a whole number of times the product $hw$ "—to quote Perrin's words.*

W. H. Bragg, in referring to Planck's theory, which enters into Bohr's theory, has said that his (Planck's) hypothesis is not so much an attempt to explain radiation phenomena as a focussing of all the difficulties into one, so that if the master difficulty is overcome a number of other difficulties melt away.

Planck's constant was originally derived from the theory that radiation from atoms takes place collectively in bundles of energy,

* *Atoms* (1916), p. 156.

so to speak, or by repeated emissions termed quanta, and that the radiated energy of frequency ι and of a homogeneous emission is $nh\nu$, $n$ being an *integer*, as stated above (see Chapter IX). In Table VII are given the evaluations of this constant as compiled by R. T. Birge.*

TABLE VII

| Method | Value of $h$ | Dependence on $e$ |
|---|---|---|
| Total radiation constant . . . . | 6·551 | $e^{4/3}$ |
| Wien's constant . . . . . . | 6·557 | $e$ |
| Theory of atomic structure . . . | 6·542 | $e^{5/3}$ |
| Einstein's photo-electric equation . . | 6·578 | $e$ |
| Quantum relation : X-rays . . . | 6·555 | $e^{4/3}$ |
| Lewis and Adam's theory of units . . | 6·560 | $e^{2}$ |
| Quantum relation : ionisation potential . | 6·579 | $e$ |

Mean value of $h = 6·554 \times 10^{-27}$ erg sec.
The original value deduced by Planck † in 1900 was $6·55 \times 10^{-27}$.

According to Bohr's theory, the energy is only radiated between the steady states, as explained above, so that if $A_1$ and $A_2$ represent the energies of the two consecutive steady states, their difference, $A_2 - A_1$, represents the energy of radiation in multiples of $h\nu$, i.e. in terms of Planck's quantum of energy, $\nu$ being the light frequency. The mean or half-value of the kinetic energy of the system is taken as $\frac{1}{2}nhw$, in which $w$ is the frequency of revolution of the electron in its orbit. The reason for the half-value is, that the electron striking into an orbit would have a potential emission energy of $nhw$, and upon reaching the final steady state of the final orbit the potential energy, so far as the radiation was concerned, would be zero, consequently the mean radiant energy would be half the maximum energy. The frequency $\nu = w/2$, which follows from the foregoing.

It must be borne in mind that the system does not radiate energy in a spontaneous manner, i.e. of its own accord as with the radio-active atoms, and therefore the surplus energy represented by the difference between $A_1$ and $A_2$ must have been acquired by virtue of some disturbance such as chemical action, or some sudden atomic convulsion impressed on the system, as occurs when passing an electric discharge through a gas which involves ionisation (see

* *Phys. Review* (1919), 14, p. 361.
† *Deutsch. Phys. Gesell.*, Verh. 2. 17, p. 237 ; or *Sci. Abs.* (1901), p. 230.

above). Thus, *absorption* of energy should *expand* the orbit, and this appears to be the case in the photo-electric effect (see Chapter XI) ; but during *radiation* the orbit is supposed to shrink.

Taking now the formula

$$3 \cdot 290 \times 10^{15} \left( \frac{1}{q^2} - \frac{1}{p^2} \right) = \nu$$

involving the Rydberg constant (see Appendix V), and making $p$ successively whole numbers 2, 3, 4, 5, 6, etc., and $q$ one of the numbers 1, 2, 3, etc., which number remains fixed for a given series of line calculations, it was possible to obtain thereby line spectra with great accuracy in the case of hydrogen and helium.

Now Bohr has evaluated the Rydberg constant in the above formula, in conformity with the theory here given, out of what might be termed the fundamental atomic constants, viz. :—

Charge of the electron in electrostatic units, $e = 4 \cdot 774 \times 10^{-10}$
Mass of the electron in grammes, . . $m = 8 \cdot 9 \times 10^{-28}$
Planck's constant in erg seconds, . . $h = 6 \cdot 554 \times 10^{-27}$

the equation being—

$$\frac{2\pi^2 \, m e^4}{h^3} = 3 \cdot 24^* \times 10^{15}$$

The agreement between the constant thus evaluated and its experimental value is within the error due to the determinations of the atomic constants given above, but a slight correction may be necessary on account of elliptic orbits (see below). The implied *modus operandi* of this evaluation is somewhat empirical, but there are a number of remarkable coincidences which give support to Bohr's theory. In general the theory has led to—

(i) The interpretation of the Balmer formula for the spectrum of hydrogen, as given above.

(ii) The evaluation of the Rydberg constant, as given above.

(iii) The determination of the normal diameter of the hydrogen atom which appears to be half the value for the molecule ($H_2$), which is $H_2 = 2 \cdot 2 \times 10^{-8}$, according to experimental evidence.

(iv) The explanation of the fact that normal hydrogen does not exhibit all the Balmer series of lines.

(v) The calculation of the lines observed in stellar spectra of hydrogen (abnormal) as well as those obtained in the laboratory (normal).

(vi) The prediction of spectral lines which were afterwards discovered (not itself a criterion, see Chapter XIX).

* Taking a slightly better value for $m$, the agreement is practically perfect.

In short, the theory has been very successful in predicting and co-ordinating certain phenomena connected with hydrogen and helium, and it appears to reveal the structure of these atoms.

Prof. A. Sommerfeld has elaborated this theory to account for the hydrogen and helium lines which are individually complex, consisting of close fine-structure lines, on the assumptions—

(i) That there are elliptical as well as circular orbits, and
(ii) That the mass of the revolving electron is not constant but suffers a change during its orbital motion.

Bohr had already suggested that, considering the hydrogen doublets, these might be treated as a relativistic effect, and Sommerfeld has constructed equations in accordance with these ideas. (See below.)

Bohr's theory, it will be seen, involves the quantum theory. Prof. A. Berthoud * makes the following clear statement which represents the general opinion held with regard to this theory: "It is known that the physical theories current during the past century, based on Maxwell's equations, do not suffice to account for the laws of radiation from black substances. To explain these laws Planck has propounded his celebrated quantum theory, according to which the energy of an oscillating body, such as an electron, can only vary in a discontinuous manner by giving up an integral number of energy quanta, each quantum being not a constant magnitude, but the product of the frequency by a universal constant," $h$ as given above. (See Chapter IX.)

In order more easily to fix mentally the above ideas, use is here made of a comparatively recent paper (1915) by Bohr † on the quantum theory and the structure of the atom, as it affords a useful summary. An abstract ‡ has appeared of which the following is nearly a full and exact quotation—one or two minor changes having been made to preserve the uniformity of style: According to the theory of E. Rutherford, in order to account for the phenomenon of the scattering of $\alpha$-rays [see Chapter VIII], the atom consists of a central positively-charged nucleus surrounded by a cluster of electrons. The nucleus is the seat of the essential part of the mass of the atom, and has linear dimensions exceedingly small compared with the distance between the electrons in the surrounding cluster. . . . While the nucleus theory has been of great utility in explaining many important properties of the atom, it is impossible by its aid to explain many other fundamental properties if we base our considerations on the ordinary electrodynamic theory ; but this can hardly be considered as a valid objection at the present time. It does not seem that there is any escape from the conclusion that it is impossible to account for the phenomena

* *Archives des Sciences* (1919), I, p. 473 ; or *Chem. News* (1920), 120, pp. 171 and 184.
† *Phil. Mag.* (1915), 30, p. 394.　　　‡ *Sci. Abs.* A (1915), p. 561.

of temperature radiation on ordinary electrodynamics, and that the modification to be introduced in this theory must be essentially equivalent to the assumptions first used by Planck in the deduction of his radiation formula, known as the quantum theory. In the present author's previous paper it was attempted to apply the main principles of this theory by introducing the following general assumptions :—

(A) An atomic system possesses a number of states in which no emission of energy radiation takes place, even if the particles are in motion relative to one another, yet such an emission is to be expected according to ordinary electrodynamics. The states are denoted as 'stationary' (non-radiant) states of the system under consideration.

(B) An emission or absorption of energy radiation will correspond to the transition between two stationary states. The radiation emitted during such a transition is homogeneous, and the frequency $\nu$ is determined by the relation

$$h\nu = A_1 - A_2 . \quad . \quad . \quad . \quad . \quad . \quad \text{(i)}$$

where $h$ is Planck's constant and $A_1$ and $A_2$ are the energies of the system in two stationary states.

(C) That the dynamical equilibrium of the system in the stationary states is governed by the ordinary laws of mechanics, while those laws do not hold for the transition from one state to another.

(D) That the various possible stationary states of a system consisting of an electron revolving round a positive nucleus are determined by the relation

$$T = \tfrac{1}{2} nhw . \quad . \quad . \quad . \quad . \quad . \quad \text{(ii)}$$

where $T$ is the mean value of the kinetic energy of the system, $w$ the frequency of revolution, and $n$ a whole number.

It may be seen that these assumptions are closely analogous to those originally used by Planck about the emission of radiation in quanta and about the relation between the frequency of an atomic resonator and its energy. It can be shown that for any system containing one electron revolving in a closed orbit, the assumption (C) together with the relation (ii) will secure a connection between the frequency calculated by (i) and that to be expected from ordinary electrodynamics, in the limit where the difference between the frequencies of revolution of the electron in successive stationary states is very small compared with the absolute value of the frequency. On the nucleus theory this occurs in the region of very slow vibrations. If the orbit of the electron is circular, the assumption (D) is equivalent to the condition that the angular momentum of the system in the stationary states is an integral multiple of

$h/2\pi$. The possible importance of the angular momentum . . . was first pointed out by J. W. Nicholson.

The present author has previously shown that the above assumptions led to an interpretation of the Balmer formula for the hydrogen spectrum, and to a determination of the Rydberg constant which was in close agreement with the measurements. The assumption (A) has recently obtained direct support by the experiments of A. Einstein and J. W. de Haas, who have succeeded in detecting and measuring a rotational mechanical effect produced when an iron bar is magnetised. These results agree very closely with those to be expected on the assumption that the magnetism of iron is due to revolving electrons and, as pointed out by Einstein and Haas, these experiments therefore indicate very strongly that electrons can revolve in atoms without emission of energy radiation.

When we try to apply assumptions analogous to (C) and (D) to systems containing more than one electron, we meet with difficulties, since in this case the application of ordinary mechanics in general does not lead to periodic orbits. An exception to this, however, occurs if the electrons are arranged in rings and revolve in circular orbits, or from simple considerations of analogy the following assumption was proposed :—

> (E) In any atomic or molecular system consisting of positive nuclei and electrons in which the nuclei are at rest relative to each other, and the electrons move in circular orbits, the angular momentum of each electron round the centre of its orbit will be equal to $h/2\pi$ in the 'normal' state of the system, i.e., the state in which the total energy is a minimum.

In a number of different cases this assumption led to results in approximate agreement with experimental facts. In general, no stable configuration in which the electrons revolve in circular orbits can exist if the problem of stability is discussed with the aid of ordinary mechanics. This is not an objection, however, since it is assumed already that the ordinary mechanics does not hold for the transition between two stationary states. Simple considerations led to the following condition of stability :—

> (F) A configuration satisfying the condition (E) is stable if the total energy of the system is less than in any neighbouring configuration satisfying the same condition of angular momentum of the electrons.

. . . The spectra emitted from systems containing only one electron are then considered, and next those from systems containing more than one electron, this being followed by a discussion of the high-frequency spectra of the elements. It is assumed in

5

these sections that the spectra considered are due to the displacement of a single electron. If several electrons should happen to
be removed from one of the rings by a violent impact, some of the
considerations adduced would not apply, since the electrons removed in this case would be replaced by those in the outer rings.
Thus there might be a rearrangement of the electrons (after the
removal of more than one of them) that would give rise to spectra
of still higher frequency than those considered.

**Sommerfeld's Extension of the Bohr Theory : and Epstein's
Studies of the Stark Effect.**—Referring to Chapter XII, it
will be seen that under the influence of a strong magnetic
field the single spectrum lines break up into various constituents. Similarly, in an electrostatic field the lines are
separated into such constituents, this latter separation being
the Stark effect. The line separations are proportional to the
field intensity.

Sommerfeld,[*] in studying the hydrogen and helium lines which
exhibit a fine-line structure with constituents, after the manner of
the Zeeman and Stark effects, has been able to account for this
complexity of the lines, which is observed without the introduction of special magnetic or electrostatic fields. Bohr had
suggested that a relativity effect might exist due to the variation
of mass of the electron with its velocity component, which would
give rise to a separation, as observed in the case of hydrogen and
helium. Each line itself appears on close examination as a series
of lines, being a sort of normally-observed spectrum concentrated
in a small space, so that, for example, the helium ' line ' $\lambda = 4686$
is split up into 6 or more lines when excited by a current passing
through a vacuum-tube with helium present. Sommerfeld therefore applies a relativity correction based upon the change in mass
of the electron due to its velocity, as represented by the equation
(see Appendix II)—

$$m_v = m_o(1 - \beta^2)^{-\frac{1}{2}}$$

for circular orbits ; and to account in detail for some of the lines
observed elliptical orbits are postulated and when their eccentricities are evaluated by a process of selection (termed " quantising ")
the agreement is exceedingly good in respect of the distribution
and intensity of the fine lines.

This treatment involves a number of interesting developments.
According to the analysis of such lines it was found that the Balmer
series contains groups of lines with the same frequency difference,
namely, 0·365 ; each respective member of the groups being
a component of a fine-line structure. This appears from experiment to be approximately the case, for Michelson has determined

* *Atombau und Spektrallinien* (1921).

the frequency difference of the Hα and Hγ lines—that is to say, the doublet differences corresponding to these lines, which are 0·32 and 0·42 respectively.   The mean value of these figures thus becomes very close to the theoretical figure ; but owing to the fine-line structure Sommerfeld has pointed out that the difference just cited should be that between *triplets*, and that the value should be 0·363.

"Paschen's careful observations"—quoting from Silberstein's *Report on the Quantum Theory of Spectra* (1920), published by A. Hilger Ltd.—" of the helium series stimulated by Sommerfeld's predictions, have not only revealed for the first time the *triplet* nature of each ' line,' but also the fine structure of each of the three components, especially for the first members of the series, $n=4$, but also $n=5$, $n=6$.   Several of the satellites, even the fainter ones, predicted by the theory, were detected, for the first time, on Paschen's photographic plates."

The following diagram shows the predicted fine lines (under high magnification) for the helium ' line ' $\lambda=4686$.   The value 4686·0 falls about midway between the first and second D-lines, counting from left to right.   At the extreme right the *a*-line

registers 4685·3.   The different styles of letters represent the calculations according to the $n$-values taken, that is—

$$3+0=A'' \ B'' \ C'' \ D''$$
$$2+1=A' \ B' \ C' \ D'$$
$$1+2=a \ b \ c \ d$$

Referring to the above difference constant (0·365) it is of theoretical interest to note that the same doublets occur in the L-series of X-ray spectra of the elements (see Chapter V).   Paschen has determined this value, viz. $0·3645 \pm 0·0045$, which is in exceedingly close agreement with the above final value 0·363, thus giving support to Sommerfeld's calculations.

Another allied phenomenon is that of the Stark effect, and Epstein[*] has been successful in applying the quantum dynamics [†] to this effect, whereby the motions of the electron in giving rise to certain lines Hβ and Hγ, as observed by Stark, are accounted for with great accuracy.   The observed and calculated values are those of $\Delta\lambda$, where (*p*) refers to the light which is polarised " so that the electric force is parallel to the lines of the external electric field, while (*s*) refers to light polarised perpendicular to this."—Jeans, *Dynamical Theory of Gases* (1921), p. 432.

* *Ann. d. Phys.* (1916), 50, p. 489 ; see also 58, p. 553 (1919).
† Not, of course, a fully-developed system of dynamics.

Z refers to an integer derived from the relations of series numbers, as used in the Bohr formula respecting the orbits.

$$H_\beta$$

| Z= | 12 | 10 | 8 | 6 | 4 | 2 | 0 |
|---|---|---|---|---|---|---|---|
| Calculated | 19·4 | 16·1 | 12·9 | 9·7 | 6·5 | 3·2 | 0 |
| $\Delta\lambda$ Observed ($p$) | 19·4 | 16·3 | 13·2 | 10·0 | 6·7 | 3·3 | 0 |
| Observed ($s$) | 19·3 | 16·4 | 13·2 | 9·7 | 6·6 | 3·4 | 0 |

The $\Delta\lambda$ values are in Angström units.

## REFERENCES

See those given in Chapter XII. R. T. Birge (*Phys. Review* (1921), 17, p. 589) makes a rigorous comparison between the experimental and theoretical results in respect of the Balmer series of hydrogen. Interesting features of the quantun theory are discussed. See Appendices IV and V.

## CHAPTER XI

### THE PHOTO-ELECTRIC EFFECT

A BRIEF statement of the photo-electric effect should be of interest, as it has been referred to elsewhere and it is closely associated with atomic theory.

When light emission takes place, it is supposed to be due to the electron striking inwards towards the positive nucleus of the atom, and according to Bohr's theory of the hydrogen atom (Chapter X) an electron jumps from an outer to an inner orbit, and during its passage monochromatic radiation is emitted.

In the photo-electric effect, the electron is caused to fly out of the atom, or out of its inter-nuclear position, by high-frequency light falling on it. Ultra-violet light incident on a clean metallic surface causes the emission of electrons from the metal. This is the photo-electric effect.

With the alkali metals, sodium and potassium, for example, visible light, i.e. light of lower frequency than the ultra-violet, and of longer wave-length, causes this electronic emission

Elaborate experiments have been carried out to get accurate and reliable measurements of this phenomenon. The surfaces must be very clean, otherwise disturbing effects arise and the measurements are not of value in gauging the effects from different substances. Non-metals give the effect also, but it is stronger with metals.

To get clean surfaces metals have been distilled in a vacuum, and in addition a species of lathe has been constructed and placed within the vacuum tube. By this means fresh surfaces could be prepared. The lathe was operated by a magnetic field passing through the sealed glass walls of the tube. By rotating the magnet which produced this field an armature fastened to the lathe spindle was rotated, and by this means the metal held on the spindle could be turned bright. The vacuum was thus maintained while the lathe was operated, being under external control.

The names of H. Hertz, W. Hallwachs, P. Lenard, A. Ll. Hughes, A. Einstein and R. A. Millikan have been particularly associated with this phenomenon. Einstein gave a simple formula which links up this effect with what might be termed the Bohr-Planck energy-equation of the atom, as exemplified in the case of hydrogen and helium, which is given in Chapter X, that is—

$$\tfrac{1}{2}mv^2 + x = hf.$$

where

> $m =$ mass of the electron.
> $v =$ maximum velocity of emission of electrons from a given metal.
> $f =$ the frequency of light that just expels the electrons with the velocity $v$ for the given metal.
> $x =$ the energy required to expel the electron from its atom.
> $h =$ Planck's constant.

Millikan determined the value of $h$ photo-electrically, with an error in the case of sodium radiation of not more than one-half of 1 per cent, the value coming out at $6 \cdot 56 \times 10^{-27}$. That is to say, using the above equation with the experimental value obtained from sodium, $h$ was found to agree with its determination by other methods (see Table VII, p. 61).

Thus the energy $hf$ absorbed by the electron and emitted in freeing itself from the atom (photo-electric effect) is equal to Planck's quantum of energy, irrespective of the frequency of the light. Therefore, light energy appears in indivisible bundles $hf$.

For a given metal or substance the effect sets in at a given light frequency. The electrons emitted in this effect are sometimes called photo-electrons, but they are the same in kind as those of cathode rays, or those given out explosively by certain radio-active atoms ; in fact there appears only to be one kind of negative electron. The word *electron* always signifies this negative one, unless otherwise stated. The positive electron has not been definitely determined or separated from the hydrogen atom or from positive portions of matter as particularly revealed in Rutherford's experiments (see Chapter VIII). There are reasons which lead to the unity or identity of the positive electron and the positive material of matter, which, in its simplest, or least aggregate form, is the hydrogen atom considered apart from its negative electron (see Chapter XX). This Chapter should be read in connection with the latter part of the next Chapter, which involves the same class of phenomenon.

## REFERENCES

The following books may be consulted with advantage :—

*Photo-Electricity*, by A. Ll. Hughes.
*Photo-Electricity*, by H. S. Allen.
*Ions, Electrons and Ionising Radiations*, by J. A. Crowther.

# CHAPTER XII

## THE ZEEMAN EFFECT: GASEOUS IONISATION: IONISATION POTENTIALS: AND MODERN VIEWS CONCERNING RADIATION

THE undulatory theory of light had been thoroughly established by mathematical and experimental analyses. Clerk Maxwell in interpreting Faraday's electrical experiments had shown that light waves must be of electromagnetic origin and, in fact, that all such waves passing through the æther could be nothing other than electromagnetic disturbances propagated with the velocity of light.

Heinrich Hertz firmly established Maxwell's theory that light was an electromagnetic-wave phenomenon. This was accomplished by producing electric waves with an oscillating electric current. These waves were of long length (see Table IV, p. 31). It was found that their velocity was the same as that of light waves; further, that they could be reflected and refracted and that it was possible to show interference phenomena; all of which indicated that they were the same fundamental phenomena as light itself.

Faraday in 1862 had tried to influence the spectrum of a sodium flame by subjecting it to the action of a magnetic field; but Faraday's spectroscope was of insufficient resolving power to produce any visible effect. Zeeman in 1896 succeeded with a powerful field and a spectroscope of greater power in broadening the spectrum lines, and upon further investigation found that when they were viewed at right angles to the lines of force (in the case of a sodium flame, for example, taking *one* of the sodium lines, as the normal spectrum gives a pair) two new lines appeared, one situated on each side of the original line. Thus *three* lines appeared where there was only *one*. When the light was viewed in a direction parallel to the magnetic lines, only the two outer lines were visible, these being circularly polarised in opposite directions to each other. This phenomenon is known as the *Zeeman effect*.

It is supposed that when an electron has its motion hastened or accelerated, radiation is given out, and H. A. Lorentz has worked out the precise system of movement necessary to give rise to this effect which, in fact, postulated the existence of negative electrons. The effect may be much more complicated than stated above, as in some cases 13 component lines are brought into existence by the magnetic field, where otherwise there would be only one line.

The effect, as is well known, is not limited to the sodium lines since those from other elements are split up by the magnetic field, and G. E. Hale has observed the effect in sun-spot spectra.

Sir G. Stokes has shown that in all probability X-rays are light waves of extremely short wave-length (wave-lengths of atomic dimensions) and that these waves have their origin in the sudden stoppage of the swiftly-moving electrons of the cathode stream. In this case acceleration of the electron takes place, and the energy given out as radiation is proportional to

$$\frac{2}{3}\frac{e^2 a^2}{c}$$

where $e$ is the negative charge of the electron, $a$ its acceleration and $c$ the velocity of light. Now, according to the Lorentz theory, as applied to the Zeeman effect, in which the magnetic field splits the lines up, at the same time shifting them relatively to the position of the original line, this action could only occur if a very minute negatively charged body of small mass were accelerated in the process. Upon resolving the components of motion necessarily involved, assuming harmonic vibrations, it was found that the results obtained agreed with the behaviour of the spectrum lines as revealed by this effect. According to this theory the shifting of the lines involves a change in the wave-length, which is proportionate to the strength of the magnetic field and the ratio of the charge of the electron to its mass. Runge and Paschen found that under the influence of a powerful field of 24,600 gauss certain mercury lines suffered clockwise ($c$) and counter-clockwise ($cc$) disturbances with reference to the original line position $\lambda_0$, with a result that the following relation has been obtained :—

$$\frac{\lambda_c - \lambda_{cc}}{\lambda_0^2} = 2 \cdot 14$$

On introducing this value into the equation, which should give the same ratio as $e/m$, a result is obtained in close agreement with the best value $1 \cdot 774 \times 10^7$ in E.M. Units. Making use of the figures given by Crowther [*] the value is $1 \cdot 65 \times 10^7$, as will be seen from the following equation which he gives :—

$$\frac{e}{m} = \frac{2 \cdot 14 \times 2\pi \times (3 \times 10^{10})}{24,600} = 1 \cdot 65 \times 10^7$$

Thus it will be seen that the electron functions in this phenomenon.

Dr. Crowther (*loc. cit.* p. 264) makes the following most interesting statement which cannot be better expressed than as stated in his own words : " It has been pointed out that a substance only emits its characteristic spectrum when it is subjected to intense ionisation. Thus we have seen . . . that the luminous parts of

---

[*] *Ions, Electrons and Ionising Radiations* (1919), p. 263.

a discharge tube are exactly those parts where the ionisation is most intense. Similarly, we have seen . . . that a flame, especially when containing salt vapours, is the seat of intense ionisation. Thus, it is only when the substance is ionised that light is emitted. It is interesting to note that Ladenburg obtained evidence of the selective absorption of the hydrogen lines when the light from a very bright discharge in hydrogen was passed through a long discharge tube containing hydrogen which was kept feebly glowing by a much weaker discharge. It would seem a fair inference that the light-emitting systems are only brought into existence when the atom has lost an electron.

" It seems probable that an electron on recombining with an ionised atom does not fall immediately into the atom but, owing to its velocity, describes a series of orbits round it, in the same way that the planets describe orbits round the sun. Each of these orbits will be described in a definite period, and will give rise to a single line in the spectrum. A single electron in passing successively through each of these orbits will thus emit in succession each of the lines in the spectrum. If the process was one of gradual collapse, so that the electron described circles of gradually diminishing radius, the light emitted would of course contain vibrations of every possible period—that is to say, it would be white light. It can, however, be shown that the number of orbits in which it is possible for the electron to move with the periodic motion necessary for the production of light is not infinite, but that their number and radii depend on the arrangement of electrons in the atom. They are thus as characteristic of the atom as the periods of its own electrons, while at the same time the light of corresponding frequency is only emitted when the atom is undergoing recombination. Bohr on the assumption that the difference in the energy of the electrons describing two successive orbits is equal to the ' quantum of energy ' . . . has succeeded in accounting numerically for some of the lines in the hydrogen spectrum." See in this connection Chapter X.

**Ionisation Potentials and Modern Views concerning Radiation.**—Although the phenomena connected with ionisation have been described in standard text-books the subject is so rapidly developing that a brief statement of the newer work and a partial survey of the subject will be given.

From the statements appearing elsewhere in these pages it will be seen that ions play a prominent part in all phenomena involving the emission of radiant energy, and the ionic state is an electrical one that is probably (1) at the basis of all chemical activity, if not representative of (2) inter-atomic forces as exemplified in certain crystals and other mass-formations of atoms. In the latter cases the condition is more a static one, whereas in the former it is more dynamic.

It is in one sense unfortunate that a single word should have extended application, since one is apt to confuse issues by not taking fully into account the precise rôle played by ions. It is hoped that this chapter will help to clear up this matter from the point of view of gaseous ionisation and radiation.

The mechanisms of ionisation and radiation are not fully known, but the latter phenomenon is believed to be a wave action that travels as an undulation (as a ripple on a pond)—as distinct from a forward compression-and-rarefaction wave (sound)—through space at a fixed speed of about 300,000,000 metres per second. Radiation wave-lengths have been measured and have been found to vary according to the type of radiation, as shown in Table IV (p. 31). In Hertz's experiments long waves, as now used by Marconi and others for wireless communication, were generated by means of an induction coil coupled to an oscillatory circuit, i.e. one containing capacity terminals, which give rise to oscillating currents of high frequency.* This oscillating circuit produced 'expanding' waves which follow one another in a sort of train fashion so that they could be treated exactly like a beam of light, in that they could be refracted by means of a huge pitch prism, reflected and focussed from large bent sheets of metal, and otherwise manipulated in practically the same way as the optician does in demonstrating various optical phenomena.†

Planck (see Chapter IX) has suggested that the atoms, or more particularly the atomic systems involving electrons, behave like oscillators and give out radiation which in the visible part of the spectrum is ordinary light.

One might broadly regard these oscillators as ionised systems which have an electric component, but their changing activity which gives rise to, or is associated with, the magnetic component is the source of radiant energy known as electromagnetic radiation. This action involves a non-steady state.

It has long been supposed that when *atoms* are ionised—that is to say, when they become active in a polar sense by the loss or gain of an electron as in salts, which are now supposed to be in a sense aggregates of ionised atoms—this ionisation may be brought about by sending an electric discharge through a tube evacuated so as to contain fewer molecules than at normal pressure. The energy emitted when they return to a more neutral or steady state appears by analysis as homogeneous radiation, as indicated by the process involved in Bohr's theory (see Chapter X).

It should be borne in mind that cases are known (1) in which radiation is emitted coincidently with the ionisation ; but in other circumstances (2) the emission seems to take place when the ions recombine. There also appear to be cases in which the radiation is not associated with (1) or (2). C. D. Child discusses these matters

---

* See Lodge, *Modern Views of Electricity* ; or Hertz, *Electric Waves*.
† See Preston, *Theory of Light*.

in the *Physical Review* (1920), 15, p. 30. The nature of the spectrum has also to be taken into account. Banded spectra are thought to be due to the action of the electrons of the molecule rather than of the atom.

Passing now to a particular kind of ionisation, ionisation by collision, this has been very thoroughly investigated by Townsend * and Thomson.† The theory of the action is fairly well developed.‡ Consider two electrodes some centimetres or millimetres apart in an enclosure termed an ionisation chamber, the electrodes may be parallel plates. When these plates are made the terminals of a source of E.M.F., a field is established which has its characteristic intensity best developed in the region between the plates. Now a few stray ions present in the gas within the enclosure will be moving about with average velocities determined by the kinetic theory of gases ; but when the field is established between the plates, some of these ions in this field will be urged towards, or away from, one plate in accordance with the electrical behaviour of such bodies. In consequence of the field these ions will have

* See J. S. Townsend, *Electricity in Gases* (1915).
† See J. J. Thomson, *Conduction of Electricity through Gases* (1906).
‡ Townsend, in his *Theory of Ionisation of Gases by Collisions*, makes the following statements, which will serve as an introduction to the subject, especially for those readers unfamiliar with this phase of ionisation. "The process of ionisation by collisions between ions and molecules of a gas may be examined by investigating the currents between parallel-plate electrodes when ultra-violet light falls on the negative electrode or when the gas is ionised by Röntgen rays. If the gas is at a high pressure, the current increases with the electric force and attains a maximum value, which is not exceeded unless very large forces are used. It is possible, however, by reducing the pressure of the gas to make the ions travel with sufficient velocity to generate others by collisions with molecules, even when the potential differences employed are small, and thus with a few hundred volts to obtain large increases in the current.

The curve, Fig. 14*a*, showing the connection between the current and electric force in a gas at low pressure, may be taken as illustrating this effect.

"In the first stage, *AB*, the current between the plates increases with the electro-motive force. The rate of increase diminishes as the force increases, and the current tends to attain a maximum value.

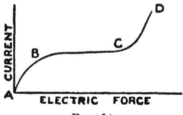

FIG. 14*a*.

"In the second stage, *BC*, the current remains practically constant and shows only small variations for large changes in the force. If the ions are produced by the action of Röntgen rays or Becquerel rays, the constant values are attained when the force is sufficiently great to collect all the positive and negative ions on the electrodes, but before this value is reached an appreciable number of ions is lost by recombination. Again, the ions may be produced by the action of ultra-violet light on the negative electrode. In this case, if the force is too small, some of the ions do not reach the positive electrode, but diffuse through the gas to the negative electrode.

"In the third stage, *CD*, when the force is further increased, there is a large increase in the conductivity. This can be explained on the hypothesis that new ions are generated by collisions, at first practically by negative ions, but as the force increases and the sparking potential is approached, the positive ions also acquire the property of producing others to an appreciable extent."

their velocities increased, and when they collide with gas molecules present the energy of the collisions causes attachments of electrons to the molecules, so that they will then possess a charge. Considering a single molecule, it thus becomes a molecular ion and its velocity may be increased, owing to the field in which it is situated, with a result that when it collides with a neutral molecule further ionisation is brought about. This action continues and the ionisation increases until a spark discharge takes place between the electrodes, provided the electric intensity between the plates is sufficient.

It should be noted that in this process of ionisation free electrons attach themselves to the molecules present in the electrostatic field, which accords with a theory developed by J. J. Thomson.

This action has been formulated, for an ion moving in an electric field of intensity $X$ acquires an amount of energy $X\varepsilon\lambda$, where $\varepsilon$ is the charge on a univalent atom (this is equal to the charge $e$ of the electron) and $\lambda$ is the free path of the ion. The mean free path of a moving particle in a gas can be calculated from the kinetic theory. In fact the velocity of an ion of mass $m$ and charge $e$ moving in a field of strength $X$ is—

$$\frac{Xe}{m} \cdot \frac{\lambda}{V}$$

where $V$ is the velocity of the particle due to thermal agitation —that is to say, the velocity considered before the field is established.

The minimum potential energy (E.M.F.) required to ionise some of the molecules of a gas involving the method of collisions, as above, has been determined for a number of gases. In this method, a metal plate (e.g. zinc) is provided which emits electrons on exposure to ultra-violet light (photo-electric effect—see Chapter XI). Parallel to this plate and a little distance from it a sheet of metal gauze is arranged (referred to as the second plate) ; it is connected with a source of E.M.F. The first plate is also connected to a source of E.M.F., but of slightly less value. A third parallel plate, which is provided with a circular hole in its centre, and arranged a little distance from the gauze, is connected to an electrometer. The ultra-violet light passes through the hole in the third plate and through the meshes of the gauze and is stopped by the first

* See, for example, L. B. Loeb, *Phys. Review* (1921), 17, p. 89, who says : " In most if not in all ionisation processes in gases the first step consists in the liberation of an electron from a neutral atom or molecule. As all evidence points to the fact that the normal negative ion in a gas is a body of molecular dimensions, the electrons formed by the ionising agent must first attach themselves to neutral gas molecules in order to form the type of carriers that are usually observed. It seems quite likely that in this process [mobility of ions at low pressures involved] may lie the explanation of the abnormal mobilities of ions observed in air at low pressures."

plate. By the action of the light, electrons emerge from the first plate and reach the second plate with a velocity, $v$, according to the equation—

$$Ve = \tfrac{1}{2}mv^2 \qquad \sqrt{\frac{2Ve}{m}} = v,$$

where V is the difference of potential between the first and second plates, $m$ is the mass of the electron and $e$ its charge (see Chapter XI). The voltage V is adjusted until the velocity of the electrons is sufficient to produce ionisation of the molecules which is indicated by the electrometer. Thus the minimum ionising potential, V, is noted, when the electrometer registers a current. As an example, the ionising potential of hydrogen is 11 volts, oxygen 9 volts, and helium 20·5 volts (see Table below for amended values).

The energy involved in effecting the ionisation of hydrogen. for example, is obtained by multiplying $4 \cdot 774 \times 10^{-10}$ electrostatic unit, the charge value for $e$, by that for 11 volts $= 1/300 \times 11$, which becomes $1 \cdot 75 \times 10^{-11}$ erg for hydrogen. Similarly, for oxygen the energy is $1 \cdot 43 \times 10^{-11}$ erg.

The above phenomenon appears to involve electron impacts which are not wholly elastic. There are at least *three* types of inelastic impact, viz. :—

(i) Impacts involving one ionisation potential, as above.
(ii) Impacts involving a single resonance potential.
(iii) Impacts involving two resonance potentials.

Resonance potential is that particular potential which corresponds to a certain line in the spectrum of the element (atom); that is to say, one subsidiary in a sense to the main spectral line— as if the electron had insufficient energy to ionise the atom, but still had sufficient energy to give effect to the particular line observed.

Nagaoka * discusses resonance spectra, and it would appear from the line of argument presented that resonance arises from forced vibrations of the charged particles of the atoms or molecules. This action, when it takes place, involves the response of all the electronic rings and nuclei of the atoms to light stimulation of extraneous origin. Whether this theory is wholly true or not it will serve a useful purpose here in calling the reader's attention to another matter of theoretical importance. R. W. Wood's experiments † on the resonance spectra of iodine vapour is considered by Nagaoka in this connection.

Returning to the phenomenon of ionisation by impacts, in a recent paper by F. L. Mohler, P. D. Foote and W. F. Meggers ‡ three types of potentials are recorded which afford values in agreement with the quantum relation $Ve = h\nu$.

* *Phys.-Math. Soc. Japan Proc.* (1919), **1**, series **3**, p. 60.
† See *Physical Optics*, 3rd edition.
‡ *Optical Soc. of Amer. Journ.* (1920), **4**, p. 364.

These types may be picked out by the notation (due to Paschen) in Table VIII.

> 1·5S is identified with the ionising potential corresponding to a limiting frequency, while
> 1·5S−$2p_2$ is identified with a resonance potential, and
> 1·5S−2P is identified with a *second* resonance potential.

### TABLE VIII

| Metal | Paschen Series Notation | Wave-length Å Units *in vacuo* | Calc. Potential | Observed Potential |
|---|---|---|---|---|
| Zinc | 1·5S | 1319·98 | 9·352 | 9·3 |
| | 1·5S − $2p_2$ | 3076·88 | 4·012 | 4·18 |
| | 1·5S − 2P | 2139·33 | 5·770 | 5·65 |
| Cadmium | 1·5S | 1378·69 | 8·954 | 9·0 |
| | 1·5S − $2p_2$ | 3262·09 | 3·784 | 3·95 |
| | 1·5S − 2P | 2288·79 | 5·393 | 5·35 |
| Mercury | 1·5S | 1187·96 | 10·391 | 10·2 |
| | 1·5S − $2p_2$ | 2537·48 | 4·865 | 4·76 |
| | 1·5S − 2P | 1849·60 | 6·674 | 6·45 |
| Magnesium | 1·5S | 1621·72 | 7·612 | 8·0 |
| | 1·5S − $2p_2$ | 4572·65 | 2·699 | 2·65 |
| | 1·5S − 2P | 2853·06 | 4·326 | 4·42 |
| Calcium | 1·5S | 2028·20 | 6·086 | 6·01 |
| | 1·5S − $2p_2$ | 6574·59 | 1·877 | 1·90 |
| | 1·5S − 2P | 4227·91 | 2·919 | 2·85 |

| | Resonance Potential | | Ionisation Potential | |
|---|---|---|---|---|
| | Calc. | Observed | Calc. | Observed |
| Nitrogen | 8·26 | 8·18 | | 16·9 |
| Oxygen | | 7·91 | | 15·5 |
| $H_2$ | | 12·22* | 16·2 | 16·5 |
| H | 10·1 | 10·4 | 13·5 | 13·3 |

This part is taken from a paper by Mohler and Foote, *Journal of the Optical Society of America* (1920), 4, p. 49.

The type of apparatus employed was in general similar to that described above, except that four electrodes were used involving a hot-wire cathode surrounded by two cylindrical grids and a cylindrical anode.

Mohler and Foote † in experimenting with phosphorus, iodine,

* Added from paper cited below.
† *Bureau of Standards Bulletin*, 16, p. 669 [Sci. Papers, No. 400] (1920).

sulphur, nitrogen, oxygen and hydrogen found, except in the case of hydrogen, that these elements exhibit resonance and ionisation potentials similar in relative magnitudes to those obtained with metallic vapours, given above. In the case of sulphur there appear to be two resonance potentials and one ionisation potential ; while hydrogen is unique in exhibiting two resonance potentials and two ionisation potentials.

Considering now another phenomenon, which is an X-ray effect, the experiments of Rutherford, J. Barns and H. Richardson * on the maximum frequency of X-rays from a Coolidge X-ray tube for different voltages, are of interest. Voltages from a Wimshurst machine were employed, these ranging from 13·2 to 142·5 kilovolts. It was found that the relation between the frequency of radiation ($\nu$) and the voltage is accurately expressed by the formula

$$h\nu = E - cE^2,$$

where $E$ is the energy of the electron, $c$ is a constant and $h$ Planck's constant, as given elsewhere. With high voltages, a correction factor has thus to be introduced into the quantum equation. According to this formula the frequency for 13·2 kilovolts would be $3·07 \times 10^{18}$. The observed frequency deduced from the absorption of the X-radiation in aluminium is $2·94 \times 10^{18}$. The corresponding values for the highest voltage, 142·5 kilovolts, are in exact agreement, being $17·4 \times 10^{18}$.

The experiments of W. Duane and F. L. Hunt,† in which they also used a Coolidge tube with a tungsten target and employed very high continuous current voltages (up to about 41,000 volts), are of special interest. The tube of course produces X-rays in abundance. It was found that the X-rays produced had different wavelengths corresponding to different glancing angles of the X-rays. From a series of measurements of wave-lengths and the voltages applied to the tube it is found that the ratio of the electron energy to the frequency of the radiation (upper limit) is represented by Planck's constant $h$. In other words, from observations of the voltage producing the ionisation current and the current passing through the tube, the value of $h$ may be deduced by the equation—

$$V_o e = h\nu_o = hc/\lambda_o$$

where $e$ = electron charge, $c$ = velocity of light, $V_o$ = the minimum voltage required to produce the X-rays of wave-length $\lambda_o$, and $\nu_o$ the limiting frequency.

The discharge through the tube is a stream of fast-moving electrons. Those in striking the anti-cathode or target generate the X-rays, and the energy so given up reappears as radiation. The energy of this radiation as measured by the electric potential

* *Phil. Mag.* (1915), 30, p. 339.       † *Phys. Review* (1915), 6, p. 166.

applied to the tube and the current (which may be evaluated from $e$) is equal to the frequency of the X-rays multiplied by $h$. This suggests a similar action in the case of the photo-electric effect (see Chapter XI) in which the ultra-violet light (corresponding in principle to X-rays) is converted into electron-emission energy involving an energy equation in which $h$ appears.

Thus it would seem that the energy phenomena are reversible, so that the radiation is as it were convertible into moving electrons and moving electrons are convertible into radiation. It is of course only the energy which is thus convertible. The mechanism of conversion is not, however, known. Millikan * says :

" The appearance of $h$ in connection with the absorption and emission of *monochromatic* light (photo-electric effect and Bohr atom) seems to demand some hitherto unknown type of absorbing and emitting mechanism within the atom. This demand is strikingly emphasised by the remarkable absorbing property of matter for X-rays, discovered by Barkla (*Phil. Mag.*, 17, p. 749), and beautifully exhibited in de Broglie's photographs opposite p. 197. It will be seen from these photographs that *the atoms of each particular substance transmit the general X-radiation up to a certain critical frequency and then absorb all radiations of higher frequency than this critical value.* The extraordinary significance of this discovery lies in the fact that it indicates that there is a type of absorption which is not due either to resonance or to free electrons. But these are the only types of absorption which are recognised in the structure of modern optics. We have as yet no way of conceiving this new type of absorption in terms of a mechanical model. There is one result, however, which seems to be definitely established by all of this experimental work. Whether the radiation is produced by the stopping of a free electron, as in Duane and Hunt's experiments, and presumably also in black-body experiments, or by the absorption and re-emission of energy by bound electrons, as in the photo-electric and spectroscopic work, Planck's $h$ seems to be always tied up in some way with the emission and absorption of energy by the electron. *$h$ may therefore be considered as one of the properties of the electron.*" The italics are the author's.

Prof. Bragg,† after describing the experiments of Duane and Hunt, says : " Exactly how this strange transfer of energy from one form to another takes place we do not know : the question is full of puzzles. The magnitudes involved are hard to realise ; it is helpful if we alter the scale of presentment. Suppose that the target of the X-ray bulb were magnified in size until it was as great as the moon's disc—that is to say, about a hundred million times— the atoms would then be spheres a centimetre or so in diameter, but the electrons would still be invisible to the naked eye. The distance from earth to moon would correspond roughly to the distance that ordinarily separates the bulb from an observer or his apparatus. We now shoot the enlarged electrons at the moon with a certain velocity ; let us say that in every second each square yard or square foot or square inch, it does not matter which, receives an electron. A radiation now starts away from the moon which immediately manifests itself (there is no other manifestation whatever) by causing electrons to spring out of bodies on which it falls. They leap out from the earth, here one and there one, from each square mile of sea or land, one a second or thereabouts. They may have various speeds ; but none exceed, though some will just reach, the velocity of the original electrons that were fired at the moon. . . ." This greatly enlarged representation, reduced again to normal size, "is the process that goes on in and about the X-ray bulb, which is part of a universal natural process going on wherever radiation, electron or wave, falls on matter, and which is clearly one of the most important and most fundamental operations in the material world."

It will be seen that in the production of X-rays the energy of the electron which is converted into radiation energy gives rise to a radiation frequency that is governed by the quantity of energy thus transferred irrespective of the atom involved, which acts as a transformer, to use Bragg's expression—and the effect is reciprocal. There is thus a limit to the frequency determined solely by the energy of the electron ; or, conversely, the movement imparted to the electrons by X-rays, or by ultra-violet light, has a corresponding limit, the process being reversible.

The constancy of the velocity of light, and the view of J. J. Thomson that lines of force play a part in the mechanism of radiation are of suggestive interest in this connection. One might raise the question : Is the constancy of these phenomena involving $h$ due to a property of space which involves lines of force that only transmit undulations when they are seized upon by electrons, or, conversely, do these lines as it were jerk out the electrons from their atom-bound positions ? One is so accustomed to think of atomic entities that the possibility of a *line* entity perhaps escapes consideration ; yet when magnetism is considered as studied by the electrician the line idea has a greater appreciation. One has, however, to think of lines of force always being present and their manifestation only occurring when they are called into action, as if they existed in space in a neutralised state.

It will be remembered that Preston in his writings * suggested that there might be vortex filaments in the æther (or of the æther) which would replace or represent lines of force. The æther was supposed to spin round these lines and, when a body is lifted against gravity, work is done in stretching the filaments. The work is thus stored as kinetic motion of the æther and, when the body returns to earth, the energy of the motion passes into the body and

* Preston's *Theory of Heat* (1919), p. 87, or earlier editions.

6

becomes the kinetic energy of its mass. Thomson's recent views of mass, energy and radiation are of interest (see Chapter XXIV).

One objection to Preston's theory was that space could hardly contain such a tangle of filaments as would result from the normal movements of masses of matter. "Moreover," to quote from Risteen, " since space vortices cannot intersect, a few seconds of intermolecular motion would suffice to tie up the vortex system into a mass of knots that would drive Gordius mad with envy, and render the ' first law of motion ' impossible."

The greatest obstacle in making use of the imagination to bridge gaps and connect experimental facts—quite a scientific and legitimate process—is to visualise properly the magnitudes involved. Rutherford's nuclear theory of the atom (see Chapter VIII) involves the conception of mass centres so small that the atom as a whole is relatively a big affair. The inter-atomic spaces, as measured from one nucleus to an adjacent one, even in a solid, is roughly in the same proportional scale as the distances between the planets (see Chapter III). There is another way of looking at these matters. If the whole starry heavens, as seen by man, could be proportionately reduced so as to be contained in an enclosure the size of a room, the millions of suns and planets would not be visible any more than the molecules in air would be visible. This perhaps gives one some idea of space magnitudes : the converse of that given by Bragg but having the same application.

The difficulty presents itself of understanding the rigidity of solid matter if it is made up of such small entities as electrons and positive nuclei widely spaced. The rigidity of a collection of nuclei and the definite bulk of atoms *en masse* are evidently due to their electrostatic fields and perhaps magnetic fields also. It is commonly believed that revolving charges (Rowland's experiment) set up magnetic lines coincident with their axes of revolution, or the axes may lie tangent to the lines of force (see Figs. 39 and 39A, pp. 115, 116). The electric conception of rigidity may be illustrated.

A basin containing small steel brads represents in a crude and greatly enlarged way a liquid ; but to improve the simile, the brads may be imagined to be small enough to suffer natural-period vibrations brought about by temperature, and consequently a certain liveliness or mobility in the mass would follow, as in the Brownian movement. Furthermore, it may be assumed that the brads are feebly magnetised, so that they would stick together slightly, but still be able to slide freely over one another.

On pouring these brads on the poles of a strongly-excited electromagnet they would be seen to cling together in a rigid manner, thus simulating the solid state of matter. It is worth noting that the liquid state is not compatible with the strongly magnetic state. There is no case on record of a liquid substance being strongly magnetic. Liquid oxygen is only feebly magnetic (see Chapter XXIII). Contact is involved in this model illustration, but contact could in a measure be avoided in principle as the forces act across air gaps, as instanced in the powerful reaction between a motor armature and its polar field—this action taking place across an air gap (or vacuum gap).

Thus it will be seen how rigid substances full of *relatively* vast spaces may exist and afford plenty of room for the free passage of lines of force without undue entanglement with electrons and positive nuclei.

Now it has been shown that the active agent in the emission of energy as radiation is the electron.  The electron is also itself a recipient of radiant energy which imparts to it the activity exhibited in the photo-electric effect.  Considering Sir J. J. Thomson's ideas, those of Sir E. Rutherford and the quantum idea as founded by Planck, one can see that there need be no tangle of lines of force though matter may be in complex movement, for the nuclei and electrons are so small that they would not more than slightly bend aside such lines if they were to pass very close to them, even if they are considered as rigid as stretched piano wires ; but when the electron does encounter a line, or by some curious affinity seizes hold of one of these wires, the energy of the electron is transferred to it and travels along the wire as a vibration or undulation.  At the other end of the wire the energy is absorbed by another electron which by some curious sort of mechanism is sent flying out of the substance of its normal abode.

These ideas are of course highly speculative, but they are helpful in showing how certain experimental facts may be linked together and better understood if they are presented to the mind by having recourse to tangible effects.

An interesting observation in connection with these ideas is that the velocity of light is always constant whatever velocity the light-source may take, and this would seem to follow if the lines of force, or by analogy the wires, can be given an undulatory movement by the electrons giving the lines a *to-and-fro lateral shake*. There would, of course, be a Doppler effect, which is a change in wave-length due to relative movement.

There are difficulties which arise in theories of this kind, one being the distribution of energy over the wave-front.

In support of the lines-of-force idea, the following taken from Jeans' Report * is of interest : " Somewhat similar ideas had been formulated before the rise of the quantum-theory, in particular by Sir J. J. Thomson (see *Camb. Phil. Soc. Proc.*, 1907, 14, p. 417), who regarded the Faraday tubes as having discrete existence in the æther.  This train of ideas is discussed by Whetham (*Experimental Electricity*, 1905, p. 207) as follows : ' Faraday's tubes, it is clear, give a powerful and convenient method of studying the phenomena of the electromagnetic field, and indications are not wanting that they represent something more than a useful mathematical fiction.

* *Report on Radiation and the Quantum Theory* (1914), p. 85.

If the structure of the electric field be discontinuous in reality, as our tube-picture of it indicates ; if the electric and magnetic effects of a charge of electricity are in reality exerted throughout the surrounding space by means of discrete tubes of force-vortex filaments in the æther, or whatever they may actually be ; an advancing wave of light must be discontinuous also. Could we look at such a wave from the front and magnify it millions of times we should see not a uniform field of illumination, but a number of bright specks scattered over a dark ground. Each tube of force would convey its own tremors and these would constitute light, but between them would lie undisturbed seas of æther.' "

Whetham remarks that such an idea about the nature of a wave-front of light (or radiation) is unexpected and surprising ; and one would be inclined to relegate such tubes or lines of force to a museum of conceptional curiosities ; but it is a remarkable circumstance that certain evidence in favour of the discontinuous nature of a wave-front really does exist. The reader may now turn to the concluding Chapter of this book. Chapter XXI is also of interest in this connection.

## REFERENCES

Zeeman, *Phil. Mag.* (1897), **43**, p. 226 ; **44**, pp. 55, 255 ; (1898), **45**, p. 197.

Zeeman, *Researches in Magneto-Optics* (1913).

R. W. Wood, *Physical Optics* (1914), p. 503.

H. A. Lorentz, *Theory of Electrons.*

Sommerfeld, *Phys. Zeit.* (1916), **17**, p. 491 ; also *Atombau und Spektrallinien* (1921), p. 416.

Debye, *Phys. Zeit.* (1916), **17**, p. 507 ; or *Göttingen Nachrichten* (1916).

Bohr, " Quantum Theory of Line-Spectra," *Danish Academy* (1918), Ser. 8, p. 79 ; see also Kramer's, same pub., p. 287.

Burgers, "Het Atoommodel van Rutherford-Bohr" (1918), Dissertation (Haarlem).

The references of later date give the advances made recently in connection with the Zeeman effect and those of the atom arising out of the work of Rutherford, Bohr and others.

# CHAPTER XIII

## THE LEWIS-LANGMUIR ATOMIC THEORY : THE OCTET THEORY : INTRODUCTION

THE theory indicated by the above title bids fair to play an important part in the elucidation of the action of atoms both from a chemical and from a physical point of view. This theory is also known as the cubic-atom theory in so far as its origin arose out of the cubical form obtained by drawing imaginary lines from regularly-spaced electrons, i.e. those arranged in cubic symmetry.

In presenting this theory, it is desirable to mention several contributory ideas that form the basis of the more fully-developed theory. In the first place, Sir J. J. Thomson prepared the way, as will be seen from the following quotation taken from his *Atomic Theory*, Romanes Lecture, 1914, p. 27, which embodies in simple language the gist of his earlier investigations : " If we assume that the recognised laws of electrical action hold for the small charges carried by the electrified parts of the atom—electrons and corresponding positive charges—we can by the aid of mathematical analysis get some idea of the way in which a number of electrons will arrange themselves when in stable equilibrium. We find that in a symmetrical atom only a limited number of such electrons can be in equilibrium when arranged on a single spherical surface concentric with the atom : the actual number which can be arranged in this way depends on the distribution of positive electricity in the inside of the atom. When the number of electrons exceeds this critical number, the electrons break up into two or more groups arranged in a series of concentric shells. This leads us to the view that the electrons in an atom, if they exceed a certain number, are divided up into groups, into a series of spherical layers, like the coatings of an onion, separated from each other by finite distances, the number of such layers depending upon the number of electrons in the atom, and thus upon its atomic weight.

" The electrons in the outside layer will be held in their places less firmly than those in the inner layers ; they are more mobile, and will arrange themselves more easily under the forces exerted upon them by other atoms. As the forces which one atom exerts on another depend on the rearrangement of the electrons in the atom, the forces which a neutral atom exerts on other atoms— what we may call the social quantities of the atom—will depend mainly on the outer belt of electrons. Now these forces are the origin of chemical affinity, and of such physical phenomena as surface tension, cohesion, intrinsic pressure, viscosity, ionising power, in fact by far the most important properties of the atom ; and the most interesting part of the atom is the outside belt of electrons. As this belt will be pulled about and distorted by the proximity of other atoms, we should expect that the properties depending on this outer layer of the electrons would not be carried unchanged by an atom through all its compounds with other elements ; they will depend upon the kind of atom with which this atom is associated in these compounds ; they will be what the chemist calls constitutive, and not intrinsic. On the other hand, the electrons in the strata near the centre of the atom will be much more

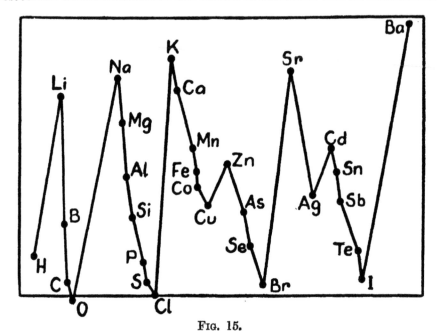

FIG. 15.

firmly held ; they will require the expenditure of much more work to remove them from the atom, and will be but little affected by the presence of other atoms, so that such properties as depend upon these inner electrons will be carried unchanged by the atom into its chemical compounds. The properties of the real atom are in accordance with these suggestions. By far the larger number of the properties of the atom are of the constitutive type which we have associated with the outer belt of electrons. There are, however, as we have seen, other properties of the atom which are intrinsic to it, these we associate with the inner layer of electrons.

" The relation between these two types of properties and the atomic weights is very different. The first type, that depending on the outer layer of electrons, waxes and wanes as we proceed along the list of elements in the order of their atomic weights ; this is illustrated by the curve in Fig. 15, which represents the variation with the atomic weight of the heat of combination of the element chlorine. The relation between an intrinsic property of the atom and its atomic weight is a much simpler one, and is of the kind shown by the curve in Fig. 16, which represents, according to the experiments of Mr. Whiddington, the relation between the energy required by cathode rays to excite the characteristic Röntgen radiation of an atom and its atomic weight; the same curve will, from the results of the experiments of Mr. Moseley and of Mr. Darwin, represent the relation between the frequency of the characteristic radiation and the atomic weight. The constitutive properties vary in a quasi-periodic and fluctuating way with the atomic weight, while the intrinsic ones steadily increase or decrease as the atomic weight increases. This is what we should have expected after our consideration of the properties of groups of electrons when in stable equilibrium. We have seen that there cannot be more than a certain number of electrons in any one layer. Consider how the atom will change as we gradually increase its population of electrons; the number in the outer layer will first increase, but when it has reached the critical number no more can be added to it; any new electrons added to the atom will now begin to form a new outer layer, the old outer layer becoming an inner one. With the addition of more electrons the same process will be repeated; the new outer layer will absorb electrons until it becomes too crowded, when a new outer layer will split off, and the process be repeated.

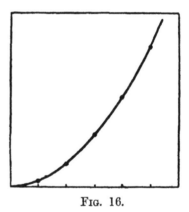

Fig. 16.

"The theory of the way in which a number of electrons arrange themselves suggests that the electrons in the atom are divided up into a series of rings, one outside the other. This has been confirmed by experiment, for the discovery by Professor Barkla of the characteristic Röntgen radiation has already enabled us to detect two of these rings in the atoms of the heavier elements and one in those of the lighter. He showed that when submitted to appropriate treatment, each atom gives out special kinds of Röntgen-rays; thus a platinum atom gives out one kind of ray, a silver atom another, with a longer wave-length than the platinum one. Now the properties of the hardest rays given out by the different elements are connected in a very simple way with the atomic weight; thus Mr. Whiddington showed that the speed of the slowest cathode particle which could excite these rays is proportional to the atomic weight, and Mr. Moseley has shown that the frequency of the vibration is proportional to the square of the atomic number ; as this number is roughly proportional to the atomic weight, the one relation would follow from the other by Planck's law. This simple connection with the atomic weight shows that these rays arise from similar parts of the atom, and the evidence is very strong that they originate in the innermost ring of electrons. Barkla has shown, moreover, that the heavier elements give out a second characteristic type of radiation very much softer than the first, which again is connected in a simple way with the atomic weight of the element.

" This radiation from elements of small atomic weights is exceedingly soft, so soft, indeed, that it has not yet been detected from any element with an atomic weight less than 90. This softer type of radiation probably originates in the second shell of electrons, counting from the inside of the atom. By the study of these radiations we thus get, in the case of the heavier elements, evidence of the existence of two groups of electrons " (see Chapter V).

Passing now to an early theory of chemical valence, Abegg and Bodländer * in 1899 suggested that every element, that is to say, the atoms of all elements, had two kinds of affinity, positive and negative, in the electric polar sense. These were termed *normal* and *contra* valence, the sum of the values irrespective of sign being 8. This hypothesis helps to explain certain chemical phenomena such as molecular compounds, in particular double salts. Association of the atoms of the same element is explained. Table IX shows the scheme arranged for the atoms of a few elements :—

TABLE IX

| Group No. . . . . . | I | II | III | IV | V | VI | VII |
|---|---|---|---|---|---|---|---|
| Elements (atoms) . . . | Li | Be | B | C | N | O | F |
| Normal Valence . . . | +1 | +2 | +3 | +4 | -3 | -2 | -1 |
| Contra Valence . . . | -7 | -6 | -5 | -4 | +5 | +6 | +7 |

Both Kossel and Lewis (see below), however, in 1916 gave a new interpretation to the conception, so that this early idea can be regarded as a forerunner to the octet theory as developed later, particularly by Lewis and Langmuir—in part, of course, since the work of J. J. Thomson, given above, afforded a very considerable contribution towards the later developments, to say nothing of the important experimental work of Moseley, which strengthened the atomic number theory of van den Broek. Mention should also be made of certain ideas advanced by A. L. Parson † in 1915, in

* *Zeit. Anorg. Chem.* (1899), 20, p. 453 ; Abegg, *ibid.* (1904), 39, p. 330.
† *Smithsonian Inst. Pub. Misc. Collections*, 65, No. 11.

particular that 8 *ring*-like electrons, called *magnetons*, after P. Weiss,* take up symmetrical positions and so represent a stable configuration. It is of interest here to note that the non-radiant states (stationary states) involved in Bohr's theory (see Chapter X) become easy of explanation if it be assumed that the electron is in reality a ring surrounding a nuclear part of the atom, since, according to electromagnetic theory, the greater the number of electrons revolving in an orbit the less the radiation due to their revolution. A ring would represent obviously a large number of electrons from the point of view of continuity ; therefore, when it is in rotation no radiation should take place—as appears to be the case. This theory of the electron, which is in effect a minute closed electric circuit with consequent magnetic properties, involves the conception of a uniform sphere of positive electricity, an idea due to J. J. Thomson and Lord Kelvin, but which does not harmonise with Rutherford's theory of a concentrated nucleus which has an experimental basis (see Chapter VIII). It is significant to note that Parson assumed that the magnetic properties of iron were due to the existence of 4 concentric shells of 8 electrons each, the electrons exerting magnetic forces on one another. In studying the crystal structure of iron by means of X-rays, A. W. Hull † found that the iron atoms are arranged according to a centred cubic lattice, each atom being surrounded by 8 others arranged in directions corresponding to the diagonals of a cube. This work of Hull appears also to be in harmony with the octet theory.

W. Kossel ‡ (March 1916), accepting the idea of electrons revolving round the nucleus of the atom in concentric orbits in one plane (see Chapter X), assigns permanent stability to the atoms of the inert gases, whilst those of other elements give up or take up electrons, these atoms striving to attain the state as represented by those of the inert gases. For example, the atoms of the alkali metals, Li, Na, etc., part with one electron per atom, leaving 8 in the outer shell, while the halogens F, Cl, etc., take up one electron per atom and thus bring the total number of outer electrons up to 8. In stereo-chemistry there is evidence that the primary valence forces between atoms act in fixed directions, and this circumstance seems to require the atoms to be arranged in a tri-dimensional scheme. This conception is therefore difficult to reconcile with the one involving electrons confined to one plane. Kossel explains the tetrahedral configuration of the valences of the carbon atom by assuming that when 4 spheres are drawn with a strong force towards a central sphere a rigid tetrahedral arrangement results : the spheres answering to the atoms. There is

---

* Weiss in 1911 used this word to designate an ultimate natural magnetic quantity or unit not involving a modification in the accepted structure of the electron.

† *Phys. Review* (1917), 9, p. 84.       ‡ *Ann. d. Physik.* (1916), 49, p. 229.

evidence that, when the carbon atoms are surrounded by fewer than 4 other atoms, the forces act in various directions, as instanced in the case of wood charcoal which has about the same volume as the original wood before it is carbonised ; and this volume may be 25 times greater than that of the carbon when in crystalline form. Thus, in both these cases the structural arrangement of the atoms— tetrahedral configuration and branching chains assumed in the case of charcoal—are difficult to conceive if the atoms are confined to orbits described round the nuclei, especially if they lie in one plane.

G. N. Lewis * (April 1916) suggested that the electrons in the atom take up fixed positions in concentric shells, each shell containing 8, except the outermost one which may accommodate 2, 4, 6 or 8 (or see Fig. 19) electrons, the last-named number repre-

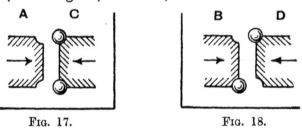

Fig. 17.                    Fig. 18.

senting the most stable combination or arrangement, as in the atoms of the inert gases, and thus form the corners of a cube— hence the name 'cubical' atom. Lewis, furthermore, suggested that when atoms combine they usually hold some of the outer electrons in common. Two electrons being so held constitute *one* chemical bond. Figs. 17 and 18 represent diagrammatically those portions of atoms about to join up in chemical combination— that is to say, A and B are about to combine with C and D to form molecules AC and BD.

Lewis gives the following diagrammatic illustration (Fig. 19) of the electrons arranged at the corners of cubes :—

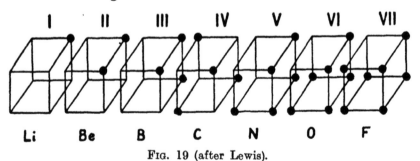

Fig. 19 (after Lewis).

I. Langmuir,† in referring to Lewis' theory, makes the following statement : " According to this theory each bond between adjacent atoms in organic compounds corresponds to a pair of electrons

* Am. Chem. Soc. Journ. (1916), 38, p. 762.
† Ibid. (1919), 41, p. 1543.

held in common by the two atoms.   Since in other types of compounds the number of pairs of electrons held in common is not
always the same as the number of valence bonds that have been
assumed, I proposed that the number of pairs of electrons which
any given atom shares with the adjacent atoms be called the covalence of that atom.   It was shown that the covalence of carbon
is always 4, that of nitrogen is usually 3 or 4, while that of oxygen
is 1, 2, or sometimes 3. . . . The results given in the tables afford
the strongest kind of evidence for the octet theory of valence
and prove that crystal form depends on the covalence of the atoms
forming the substance rather than upon the valences given by the
ordinary theory.''

Born and Landé*(1918) have also come to the conclusion that the
atom has a cubic symmetry ; the evidence in this case being largely
based upon the compressibility of the salts of the alkali metals :
e.g. NaCl.   It was suggested later that the electrons in the inert-
gas atoms and those of the halogens and the alkalis were arranged
at the corners of a cube where the electrons may describe orbits
and yet preserve the cubic symmetry of the atom.

Langmuir † in developing the ideas sketched above, together
with new material, has apparently succeeded in co-ordinating the
entities of the atom, particularly its electrons, in such a way as to
be able to account for many diverse phenomena ; one of the main
features of Langmuir's theory being the formula which serves to
characterise the molecular compound as well as the ionic state.
This theory will be given in the next chapter, which is a continuation
of this introductory part.   Later chapters will give certain developments of the theory.

There are several theories of valence, such as those of Sir Wm.
Ramsay and J. Stark, which may be regarded as pioneer work in
this field.   These will be referred to later.

* *Verh. d. Phys. Ges.* (1918), 20, p. 210.
† *Am. Chem. Soc. Journ.* (1919), 41, pp. 868, 1543.

# CHAPTER XIV

## THE LEWIS-LANGMUIR ATOMIC THEORY: THE OCTET THEORY: ARRANGEMENT OF ELECTRONS IN ATOMS AND MOLECULES *—continued

In developing this theory the main guiding principle seems to have been the *atomic numbers* of the *inert gases*, and the well-known fact that these gases are chemically inactive, as their name implies.

The atomic numbers are known to be closely, if not exactly, related to the number of negative electrons in the atom ; and in this case of the inert gases there may be perfect identity in this respect. The disposition of the electrons in the atoms of the inert gases should therefore afford a criterion for stability,† since these elements are without chemical activity. The atomic numbers are : He=2, Ne=10, Ar=18, Kr=36, Xe=54, and Ems.=86. The last members of this group being the radio-active emanations, they are taken collectively as *one* element, being probably isotopic. It is intended that the abbreviation (Ems.) should signify this fact.

Examining these atomic numbers mathematically, J. R. Rydberg ‡ pointed out that they could be obtained from the following series formula, in which the number A has successive values as the squared figures, or terms of the equation, are added, thus—

$$A=2(1+2^2+2^2+3^2+3^2+4^2 \ . \ . \ .)$$

This equation suggests symmetry (see Table XA below) and the pairing of electrons, since the same term occurs twice. Taking into account the foregoing ideas (see Chapter XIII), Langmuir has drawn up a specification containing 11 postulates which, together with a modified periodic table and a simple equation, seems to throw a great deal of light upon the problem of atomic constitution.

In these postulates it should be understood that for the most part the terms used are for the purpose of defining the positions of electrons in space. Whether they revolve or oscillate round these

---

* In addition to the literature cited in the previous chapter, the following papers should be noted : Langmuir : *Frank. Inst. Journ.* (1919), 187, p. 359 ; *Am. Chem. Soc. Journ.* (1919), 41, pp. 868, 1543 ; (1920), 42, p. 274 ; *Proc. Nat. Acad. Sci.* (1919), 5, p. 252 ; *Indus. and Eng. Chem. Journ.* (April 1920) ; *Chem. News* (1920), 121, p. 29.

† It is to be noted that the term *stable* is used in a fundamental sense.

‡ See *Phil. Mag.* (1914), 28, p. 144, and references therein.

positions is not decided by this theory. Explanatory statements will be given after some of the postulates, and diagrams will be introduced where it is considered desirable. The postulates proper are designated by the numbered paragraphs.

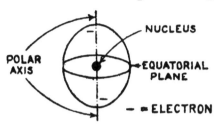

FIG. 20.

1. The electrons in atoms occupy definite positions. In the most stable atoms, those of the inert gases, the positions of the electrons are symmetrical with respect to an equatorial plane passing through the central part of the atom (Fig. 20). No electrons are situated in this plane. Perpendicular to this plane is a polar axis through which four equally-spaced planes pass, the angles between adjacent planes being 45° (Fig. 21). The symmetry thus obtained can be compared with that of a tetragonal crystal.

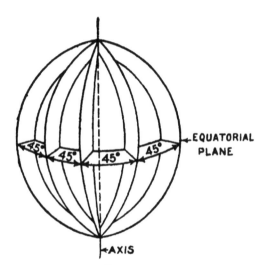

FIG. 21.

2. The electrons in any given atom are distributed through a series of concentric (or nearly concentric) spherical shells, all of the same thickness, so that their radii will form an arithmetical series : 1, 2, 3, 4 ; whilst the corresponding areas of the shells are in the ratios : $1^2 : 2^2 : 3^2 : 4^2$ (Fig. 22).

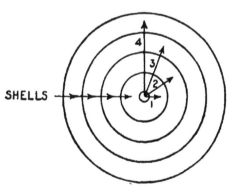

FIG. 22.

3. Each shell is divided into cells of equal area and symmetrically disposed in accordance with postulate 1. The volume of each cell in a given atom * is the same. The first shell has 2 cells, the second

* The expression *a given atom* probably means an atom under given conditions. For example, in a free atom of sodium the cells may be of greater volume than those of a similar atom when combined with some other atom.

8, the third 18 and the fourth 32. Considering hemispheres, the numbers of cells in successive half-shells are 1, 4, 9, 16.

This arrangement of cells and electrons has the symmetry indicated by Table X.

TABLE X

| Shell | Radius | Cells per Hemisphere | Cells axially placed | Cells in Zones |
|-------|--------|----------------------|----------------------|----------------|
| I | 1 | 1 | 1 | 0 |
| II | 2 | 4 | 0 | 4 |
| III | 3 | 9 | 1 | 8 |
| IV | 4 | 16 | 0 | 16 |

It will be seen that the cells in zones are always a multiple of 4. This suggests tetragonal symmetry for the atoms of the inert gases.

4. The innermost part of the atom having two cells (see Fig. 23) can contain only one electron each, but every one of the shells exterior to this part can contain cells holding, as a maximum number, two electrons each. Langmuir here remarks : " If, as Rydberg * believes, there are two undiscovered elements of atomic weights less than that of hydrogen, then this exception in the case of the innermost cells (only two) may be avoided." See in this connection Chapter XX, where an argument is presented that hydrogen may be complex and, apart from its electron, made up of a fractional and a whole-number part. Returning to the postulate, the inner cells must contain their maximum quota of electrons before the outer cells can contain any. In the outside shell two electrons can occupy a single cell only when all the other cells contain at least one electron. It is assumed that when two electrons occupy the same cell they are disposed at different distances from the centre of the atom. Each cell when containing its full quota of electrons (2) is divisible into *two layers* which may be designated 1st and 2nd, or *a* and *b*.

From the above postulates it is now possible to set out a series of figures representing the arrangement of electrons in the atoms of the inert gases ; but, before coming to these, Table Xᴀ should

* *Phil. Mag.* (1914), 28, p. 144.

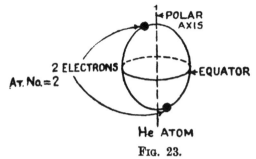

FIG. 23.

I = 1st shell; II = 2nd shell; III = 3rd shell, etc.   a = 1st layer; b = 2nd layer.

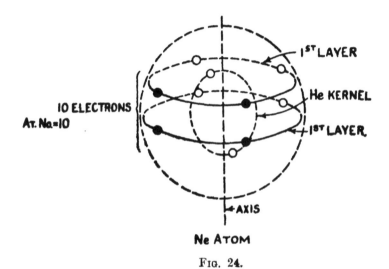

Ne ATOM

FIG. 24.

Ar ATOM

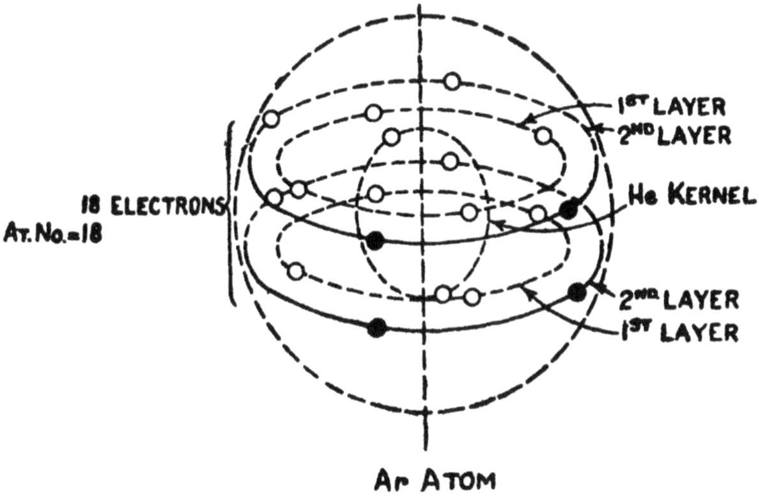

FIG. 25.

be studied, as it shows the numerical relations which form an important part of this theory.

TABLE Xa

| Shells and Layers involved | Electrons | Cells | Atom and its Atomic No. | Rydberg Atomic Number Series |
|---|---|---|---|---|
| 1st shell | $2 = 2 \times 1^2$ } 2 cells | | He = 2 | 0 } 2 |
| 2nd shell 1st layer (a) | $8 = 2 \times 2^2$ } 8 cells | | Ne = 10 | 2 } 10 ; 8 } 18 |
| 2nd shell 2nd layer (b) | $8 = 2 \times 2^2$ | | Ar = 18 | 8 } 36 |
| 3rd shell 1st layer (a) | $18 = 2 \times 3^2$ } 18 cells | | Kr = 36 | 18 } 54 |
| 3rd shell 2nd layer (b) | $18 = 2 \times 3^2$ | | Xe = 54 | 18 } 86 |
| 4th shell 1st layer (a) | $32 = 2 \times 4^2$ } 32 cells | | Ems. = 86 | 32 |
| 4th shell 2nd layer (b) | | | | |

Langmuir says that the existence of cells independent of the electrons in them seems necessary to account for the properties of the elements beyond those of the rare-earths; and herein lies a possible relation to the 'stationary' states in Bohr's theory which involves orbital radii related to one another by the series: $1^2 : 2^2 : 3^2 : 4^2$ . . . (see Chapter X); and, moreover, the passage of electrons from cell to cell, instead of from orbit to orbit, may account for line spectra. Langmuir also sees in the twofold structure of the Rydberg formula given above, a connection with the occurrence of electrons in pairs, as would be the case when two electrons occupy one cell. If, however, hydrogen is complex (see Chapter XX), following upon Langmuir's suggestion, a number of points arise; such, for example, as the initial atomic number of the series (other than unity for hydrogen), and the Bohr-Rutherford theory of the atom in relation to such a change. The present-day opinion does not seem to favour a complexity of this type.

It is of possible interest here to note that Sommerfeld's refinement of the Bohr theory (see Chapter X) which seems adequately to account for the fine-line spectrum of hydrogen and also that of helium, involves the conception of elliptical orbits. These are *quantised* (a term used by Sommerfeld to indicate a process of working out, or selecting, the eccentricities of the orbits) with the introduction of relativity dynamics. The mathematical treatment involved in respect of such differential quantum 'cells,' or quantised elliptical orbits of step-wise eccentricities, may possibly find an interpretation in, or fit into, the type of atom hinted at in the foregoing paragraph ; but it is not possible to say very much in the absence of experimental support to the idea of a complex hydrogen atom in the sense here implied. It is perhaps of significance to observe that the line spectrum of hydrogen is fairly complex. R. W. Wood has actually extended the Balmer series of hydrogen lines in the laboratory, as indicated in the early part of Chapter X. See also Appendix IV.

5. Two electrons in the same cell do not repel nor attract each other with strong forces. The outer electrons tend to line themselves up radially with the inner ones, while the electrons in the outer layer tend to distribute themselves uniformly amongst the available cells. The equilibrium positions of the electrons depend upon a balance between three forces acting in common with them, together with a mutual attractive force between the electrons and the kernel. Parson's magneton theory involving magnetic force, distinct from electrostatic force, is mentioned in this connection, which force is thought to counteract that of electrostatic repulsion.

With several forces in evidence one can account for any spacial arrangement of the electrons ; and, of course, the precise evaluation and distribution of such forces are unknown ; but there is strong evidence of their existence in the atom. Rutherford's work on the atomic structure is of importance in this connection.

6. When the number of outer electrons is small, the configuration is determined by the attraction (presumably magnetic) of the underlying electrons and, as the number of outer electrons increases, electrostatic repulsion comes into play. "As a result, when there are few electrons in the outer layer these arrange themselves in the cells over those of the underlying shell, but where the outside layer begins to approach its full quota of electrons the cells over the underlying electrons tend to remain empty."

These complexities have to be considered in the light of resulting actions in chemical phenomena, and they need not be commented upon here.

7. The properties of the atoms are determined principally by the number and arrangement of electrons in the outside shell and by the ease with which the atom is able to revert to a more stable form. This action involves one or the other of two operations, viz. (1) *giving up electrons*, and (2) *taking up electrons*—from the point of view of a given atom losing or gaining electrons.

The Periodic Tables to follow will help to give expression to the foregoing postulates.

8. The stable and symmetrical arrangement of electrons is completely exemplified in the atoms of the inert gases, as stability

7

in the disposition of the electrons goes hand in hand with their symmetrical arrangement; and these atoms are characterised by their strong internal and weak external electric fields. The external field of the atom weakens as its atomic number becomes less.

Werner * had suggested that the distribution of forces round the atom governs its stability. It would be possible to cite other suggestions of historical interest which bear upon this idea (see below).

9. The most stable arrangement of electrons is that of the *pair* in the helium atom. A stable pair of electrons also may be held by—

    (i) A single hydrogen nucleus.
    (ii) Two hydrogen nuclei.
    (iii) The hydrogen nucleus and the kernel of another atom.
    (iv) Two atomic kernels—but this combination is very rare.

10. The next most stable arrangement of electrons is the *octet*, which is a group of 8 electrons, like that of the second shell of the neon atom (see Fig. 24). Any atom with an atomic number less than 20, and which has more than 3 electrons in its outside layer, tends to take up enough electrons to complete its octet.

11. The octets may hold one, two or sometimes three pairs of electrons in common. One octet may share 1, 2, 3 or 4 pairs of its electrons with 1, 2, 3 or 4 other octets. One or more pairs of electrons in an octet may be shared by the corresponding number of hydrogen nuclei. No single electron can be shared by more than two octets.

The principle of *sharing* electrons in a rather definite way was proposed by J. Stark,† who suggested that when chemical combination occurs between two atoms the attraction takes place through an intermediate electron which forms a bond between the atoms, as illustrated by Fig. 26 : this being the simplest form of chemical union. Atoms also have 'positive patches' (+), as indicated by Fig. 27. These patches are able to bind other atoms by lines of force extending to the electrons of other atoms, as shown by Fig. 28. In this case the number of lines per electron is fewer. These figures are similar to those given by Campbell (*loc. cit.*).

It will be seen that the present theory involves electrons 'going over' entirely from one atom to another in certain cases, but in others the electrons are shared in much the same way as suggested by Stark. Sir Wm. Ramsay's ‡ theory is also of interest in this

---

* See Cohen's *Organic Chemistry* (1918), I, pp. 60, 82, 85 and 90 ; also Werner, *New Ideas in Inorganic Chemistry* (1911).
† See Campbell, *Modern Electrical Theory* (1913), p. 341.
‡ *Chem. Soc. Trans.* (1908), 93, p. 774. See also J. N. Friend, *Theory of Valency.*

connection for, according to Ramsay, "electrons are atoms of the chemical element, electricity ; they possess mass ; they form compounds with other elements ; they are known in the free state, that is, as molecules ; they serve as the ' bonds of union ' between atom and atom." Furthermore, in ionised substances, such as sodium chloride dissolved in water, the sodium parts with its electron while the chlorine atom takes over the electron given up by the sodium atom. It will be seen that, according to Langmuir's theory and the work of Milner, Ghosh and Bjerrum, * the ionisation takes place in the solid sodium chloride, for example, so that here too there is a difference between the two theories.

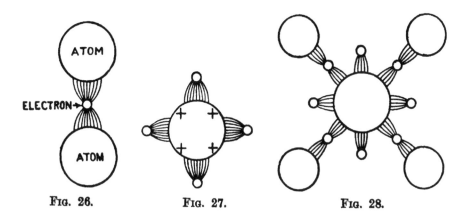

FIG. 26.                    FIG. 27.                    FIG. 28.

It is claimed by Langmuir that this theory explains—

    (i) The periodic properties of the elements, including those exhibited by the members of the eighth and rare-earth groups.

    (ii) The magnetic properties of the elements.

    (iii) The physical properties of the elements generally, such as the boiling-points, electrical conductivities, etc.

    (iv) The chemical properties as distinct from the foregoing, involving valence, ionisation (ions), etc.

The theory might be described as a specification of the elements in terms of electrons in the atom in conjunction with positive nuclei whereby the molecular and ionic conditions of matter are systematically accounted for. The positive nuclei are not shown in the diagrams (except Figs. 20, 48 and 49), but their presence in association with the electrons is implied. A brief statement of the working of the theory will be given here, followed by a more detailed treatment.

Under the influence of the mutual attraction between the atomic nucleus, or nuclei, and outlying electrons, together with the mutual

* See Chapter XV.

repulsions between the electrons themselves, a condition is established whereby electrons tend to form into definite groups or configurations. This reminds one of the early views of J. J. Thomson * and Nagaoka,† to cite two of the pioneer contributors to the subject of atomic constitution of the Saturnian type (see Chapter X).

In the first shell the stable formation is the pair. In subsequent shells, with their attendant layers of electrons, the most favourable disposition is that of the octet, so that the actual arrangement of the electrons in space is at the eight corners of a cube. The electrons may revolve about these corners, but this detail, though important, need not be considered here.

If, therefore, the electrons in a given atom are unable themselves to form a permanent or stable configuration, they may do so when brought into close association with other atoms deficient in stability ; and thus by mutual association the atoms in question tend to form a completed octet by common or joint adjustment of the electrons in the respective atoms, this being the most stable arrangement other than the pair. *This mutual action affords an explanation of one phase of chemical combination.*

As already stated, the inert gases comprise atoms of maximum stability, since they form no compounds with other atoms, nor do they associate or combine amongst themselves to form simple molecules. The inert atoms form the starting-points for formulating various degrees of less atomic stability. Thus helium may be taken as a starting-point in building up less stable atoms up to neon. Then neon becomes the starting-point for building up less stable atoms up to argon, and so on. These inactive gases, therefore, form the turning-points in the periodic classification of the elements ; or, stating this in another way, all the elements lying between those of the inert group represent in a sense a series of transitory types that attain to perfect stability (inertness) only when they have internally completed octets involving filled cells, as in the atoms of the inert elements. Consequently, these intermediate types are free to form themselves into various more or less stable compounds ; thus striving to attain more or less the perfectly inert type when they come into close contact with each other, and in so doing they form octets either *jointly* or *severally*, as will be seen from what follows.

It is necessary, therefore, to pay particular attention to the structure of the atoms of the inert elements. The above statements will be better understood by referring to Figs. 23, 24 and 25, and postulates 3 and 4. In considering the elements falling between the inert gases the following Periodic Tables should be studied.

---

* *Phil. Mag.* (1904), 7, p. 237.
† *Nature* (1904), 69, p. 392 ; see also *Math. and Phys. Soc. Tokio* (1905), 2. 20, p. 316.

Table XI is the one Langmuir gives in his original paper.* See in this connection postulate 4 and Table XA. The accompany-

TABLE XI (Langmuir)

| SHELL | LAYER | A | B→0 | 1 | 2 | 3. | 4 | 5 | 6 | 7 | 8 | 9 | 10 |
|---|---|---|---|---|---|---|---|---|---|---|---|---|---|
| I | | | | H | He | | | | | | | | |
| II | a | 2 | He | Li | Be | B | C | N | O | F | Ne | | |
| II | b | 10 | Ne | Na | Mg | Al | Si | P | S | Cl | Ar | | |
| III | a | 18 | Ar | K | Ca | Sc | Ti | V | Cr | Mn | Fe | Co | Ni | |
| | | | | 11 | 12 | 13 | 14 | 15 | 16 | 17 | 18 | | |
| III | a | 28 | Niβ | Cu | Zn | Ga | Ge | As | Se | Br | Kr | | |
| III | b | 36 | Kr | Rb | Sr | Yt | Zr | Nb | Mo | 43 | Ru | Rh | Pd | |
| | | | | 11 | 12 | 13 | 14 | 15 | 16 | 17 | 18 | | |
| III | b | 46 | Pdβ | Ag | Cd | In | Sn | Sb | Te | I | Xe | | |
| IV | a | 54 | Xe | Cs | Ba | La | Ce | Pr | Nd | 61 | Sa | Eu | Gd | |
| | | | | 11 | 12 | 13 | 14 | 15 | 16 | 17 | 18 | | |
| IV | a | | | Tb | Ho | Dy | Er | Tm | $Tm_2$ | Yb | Lu | | |
| | | 14 | | 15 | 16 | 17 | 18 | 19 | 20 | 21 | 22 | 23 | 24 |
| IV | a | 68 | Erβ | Tmβ | $Tm_2β$ | Ybβ | Luβ | Ta | W | 75 | Os | Ir | Pt | |
| | | | | 25 | 26 | 27 | 28 | 29 | 30 | 31 | 32 | | |
| IV | a | 78 | Pt | Au | Hg | Tl | Pb | Bi | RaF | 85 | Nt | | |
| IV | b | 86 | Nt | 87 | Ra | Ac | Th | $Ux_2$ | U | | | | |

A = number of electrons in the part inside the outer shell or outer shell-layer, as a shell-layer begins to form at each inert gas: see arrow in Fig. 29.

B = number of electrons in the outer layer of the atom—the chemically active part.

ing Fig. 29 exemplifies the shell idea with its layers surrounding the inner part of the atom which is referred to collectively as the

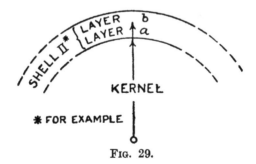

FIG. 29.

'kernel,' though it would itself contain one or more shells and layers. The kernel might be defined as the inner part of the atom which is chemically inactive towards other atoms.

* Am. Chem. Soc. Journ. (1919), 41, particular page 874.

TABLE XII is a modification of Langmuir's which is designed with a view to bringing into prominence certain features, viz. :—

## TABLE XII

CHEMICAL →  SUMMATION →

| Layer | Shell | A+B | 0 | 1 | 2 | 3 | 4 | 5 | 6 | 7 | 8 | 9 | 10 | 11 | 12 | 13 | 14 | 15 | 16 | 17 | 18 |
|---|---|---|---|---|---|---|---|---|---|---|---|---|---|---|---|---|---|---|---|---|---|---|
| I? | | | * | | | | | | | | | | | | | | | | | | |
| I | a | 2 | He | Li | Be | B | C | N | O | F | | | | | | | | | | | (Ne) |
| II | b | 10 | Ne | Na | Mg | Al | Si | P | S | Cl | | | | | | | | | | | (Ar) |
| III | a | 18 | Ar | K | Ca | Sc | Ti | V | Cr | Mn | Fe | Co | Ni | Cu | Zn | Ga | Ge | As | Se | Br | (Kr) |
| III | b | 36 | Kr | Rb | Sr | Yt | Zr | Nb 19 | Mo 20 | | Ru 22 | Rh 23 | Pd 24 | Ag 25 | Cd 26 | In 27 | Sn 28 | Sb 29 | Te 30 | I | (Xe) 32 |
| IV | a | 54 | Xe | Cs | Ba | La | Ce | † | W | | Os | Ir | Pt | Au | Hg | Tl | Pb | Bi | {RaF} ‡ | | (Ems) |
| IV | b | 86 | Ems ‡ | | Ra | Ac | Th | Ux2 | U | | | | | | | | | | | | |

Summation: 6 6 6 6 6 6 | 6

* = 0 or l.    A + B = Atomic number.    At. No. of U = 92 (H = 1).

† Continued from cerium→  5 6 7 8 9 10 11 12 13 14 15 16 17 18  Pr Nd, —, Sa, Eu, Gd, Tb, Ho, Dy, Er, Tm, Tm2, Yb, Lu.

{ } = Radio-active series, which overlaps to Tl when all the products are inserted, only leading ones being here shown. The table is complete in respect of places, however, as the radio-atoms not shown would be isotopic with those shown and fit into the places occupied by those elements shown from Tl to U.

‡ No radio-atoms have been discovered which are eligible for these places.

NOTES.—Since the above Table was prepared, a revision in respect of some of the rare-earth elements becomes necessary; so that the above series is now Pr 5, Nd 6, —7, Sa 8, Eu 9, Gd 10, Tb 11, Dy (sometimes written Ds) 12, Ho 13, Er 14, Tm 15, Yb 16, Lu 17, Ct 18. In order to bring this Table into better order chemically, a modification is given opposite the title-page as a frontispiece. However, the above Table may still be retained as a more condensed form; or see Langmuir's even more condensed Table on page 101, should this type be preferred.

(i) That there may be breaks in the step-wise continuity of the elements whilst the atomic numbers form a continuous sequence. See page 155.

(ii) That hydrogen may be complex and consist of a whole-number part $a$ plus a fractional part $b$, these hypothetical parts being assigned positions in the table. The electron would, of course, form the *third* entity.

These matters are discussed in Chapter XX, but it was thought best to place the two tables near to each other. It is not, however, suggested that one table is better than the other. There does not appear at present to be any finality about periodic tables, each one having certain merits according to the point of view taken. It is suggested that the latter table brings into prominence the significance of the A-values, especially in respect of symmetry.

NOTES.—The reason why argon, for instance, is inert, whereas iron is a somewhat active element, is apparent from the supposition that the former contains a completed stable octet of electrons in its outer layer ; whilst iron, though having 8 electrons in its outer shell, has room for more.

It will be seen that some of the radio-active elements are included in the above tables. For a complete and eminently satisfactory classification of the radio-active elements, see Soddy's monograph, *Chemistry of the Radio-Elements* (1914), Part II (or later edition with Parts I and II combined in one volume) ; or Soddy, *Chemical News* (1913), 108, p. 168 ; or *Annual Reports of the Chemical Society* (London, 1918), p. 200. See Appendix I.

Soddy's Table would fit in the bracketed parts where only a few members are shown, not to complicate the table unduly.

For a list of the Elements, Atomic Weights, etc., see Appendix I.

While there is lead which is the end-product of radio-active change, there is the ordinary lead which appears to be, together with bismuth, the stable end-product of *evolution*, as distinct from that of *devolution* characterising the radio-active process. Thus there appear to be leads of two main origins, but they both occupy the same place in the table. They are thus isotopic. These leads have appreciably different atomic weights, but their atomic number is the same, being 82 (54+28). There appear to be at least *four* leads, viz. :—

(i) Ordinary lead—a product of evolution.

(ii) Thorium lead—a product of devolution, having its starting-point in thorium. As there is a branch line of descent after thorium emanation there would be two end-products, probably both lead.

(iii) Radium lead—also a product of devolution, but having its origin in uranium, with a line of descent through radium. There is a branch in this series after the emanation ; hence two end-products exist.

(iv) Actinium lead—another product of devolution having its origin in uranium (?), but with a line of descent through actinium.

Thus it will be seen that there are a number of leads and these differ more or less in atomic weight judging from the atomic weights which have been accurately determined in the case of leads from different minerals. For example, a mineral rich in thorium would yield a lead differing in atomic weight from a lead obtained from a mineral rich in uranium or radium. The theoretical values, deduced from the a-particles given out in the process of disintegration, agree fairly well with the experimental values. It cannot be positively asserted that every one of these end-products is lead, but the evidence is accumulating to the effect that most of them are lead, being isotopic with ordinary lead. See Appendix III.

Considering hydrogen in the *atomic condition* it has one electron, and being unsaturated (see postulates 1 and 7) tends to take up an electron and assume the symmetrical form characteristic of helium. H therefore has a valence of unity. H probably has an atomic number of unity and, while it is very active as an atom or ion, when in the molecular state ($H_2$) it is very stable, like helium, since the two electrons of the pair are shared equally by the hydrogen nuclei. According to Table XI, $A=0$ and $B=1$. The atom is active chemically because it easily gives up its electron to another atom, whereby a stable pair is formed. When $H_2$ is formed the two electrons involved form a stable pair, according to postulate 9ii. This molecule has a particularly weak external field (postulate 8), and in consequence its boiling-point is low and it is relatively inert.

Considering helium, in this atom the first shell is completed. Beyond this point the additional electrons form a part of the first layer of a second shell (II$a$). It will be seen that $A=2$ and $B=0$. This atom has the stable pair already formed, and like $H_2$ it has a weak stray field and consequently forms no chemical compounds, while it has the lowest boiling-point and highest ionising potential of any known element, which facts are in accordance with the extreme weakness of its external field (see postulate 8). Molecular hydrogen, on the other hand, has a higher boiling-point and lower ionising potential on account of the repulsion between its two nuclei, which effects sufficient separation to give rise to an external field that is not negligible, as in the case of helium.

Considering lithium ($A+B=3=$ At. No., while $B=1$), in this atom the first layer or kernel is complete and therefore stable, by analogy with the inert gases, while the single electron in the outer layer gives rise to chemical activity. Lithium easily parts with its electron, leaving the atom as a univalent cation. Fluorine, which is at the other end of the series, has an atomic number of 9 and requires but one electron to complete its octet. When these two elements are brought together, a condition of stability is easily fulfilled, since the lithium atom can give up its electron to the fluorine atom and thereby cause two octets to be established; but in this case an electrostatic force is evident which draws the two atoms together (when considering a pair), since the lithium kernel would be positively charged and the fluorine atom negatively charged. This action gives rise to the ionic state and in solids—lithium fluoride in this case—the atoms would arrange themselves in a space lattice, as revealed by X-ray analysis of such salts. In the liquid state and in aqueous solutions the lithium and fluorine ions would be free to move about. Melted lithium fluoride would afford free electrons which form the mechanism of electric conduction. Similarly, aqueous solutions conduct on account of the mobility of the ions—that is to say, the state of freedom thus produced allows a free progressive movement of the electrons which constitutes the electric current.

It will be seen that in this process molecular compounds are not formed, and a solid mass of lithium contains atoms with sufficiently intense stray electric fields to give rise to cohesion ; the very great difference between the boiling-point of helium and that of lithium is thus explained.    In the vapour state lithium is monatomic owing to the thermal agitation overcoming the electrostatic forces which bind the atoms (ions) together in the solid state.    In this state, pure lithium would consist of nuclei (ions) and electrons probably arranged in a space lattice similar to the lithium and fluorine ions in lithium fluoride.

When lithium and hydrogen atoms come together the electrons of each atom form stable pairs with the hydrogen nuclei as centres (see postulate 9i) ;    the lithium kernels become positive ions ($Li^+$), as in the case above, and the hydrogen nuclei being surrounded by pairs of electrons become negative ions ($H^-$).    " These charged particles would be attracted to each other, but since there is no tendency for negative *ions* to form pairs about positive kernels there would be no tendency to form molecules.    The lithium and hydrogen ions form a crystalline solid having the composition LiH.    Since there are no free electrons, the solid body is a non-conductor of electricity.    If melted, however, the positively and negatively charged particles should be able to move under the influence of an electric field so that molten LiH should be an electrolyte (as Lewis has pointed out) in which hydrogen should appear at the anode.    The comparative ease with which an electron can be taken from a lithium atom by an electronegative element makes univalent lithium ions stable in water solutions."

Sodium and fluorine afford a parallel case to that of lithium and fluorine.    Sodium has one more electron than is needed for a condition of maximum stability and fluorine has one electron too few. When these atoms come together the extra electron of sodium passes over completely to the fluorine atom thereby completing its octet and binding sodium at the same time by electrostatic forces, as one becomes as strongly positive as the other is strongly negative—that is, they become so to each other owing to the circumstances involved.

In the case of sodium and lithium it will be seen that they give up their electrons.    This is *positive* valence=1.

On the other hand, fluorine takes up an extra electron to complete its octet.    This is *negative* valence=1.

There is a third type of valence, called covalence, which is the number of pairs of electrons an atom can share with its neighbours. This will be elucidated later.

To fix the above ideas more firmly in the mind, the following quotations from one of Langmuir's papers * should be of particular interest :    " It is obvious then that the extra electron of the sodium atom should pass over *completely* to the fluorine atom.    This leaves the sodium atom with a single positive charge while the fluorine atom becomes negatively charged.    If the two charged atoms or

* *Journ. of Industrial and Eng. Chem.* (April 1920).

ions were alone in space they would be drawn together by the electrostatic force and would move as a unit and thus constitute a molecule. However, if other sodium and fluorine ions are brought into contact with the 'molecule' they will be attached as well, as the first one was. There will result—at not too high a temperature—a space lattice consisting of alternate positive and negative ions and the 'molecule' of sodium fluoride will have disappeared. Now this is just the structure which we find experimentally for sodium by Bragg's method of X-ray crystal analysis. There are no bonds linking individual pairs of atoms together. The salt is an electrolytic conductor only in so far as its ions are free to move. In the molten condition or when dissolved in water therefore it becomes a good conductor."

Magnesium with an atomic number of 12 combined with oxygen of atomic number 8 forming magnesium oxide, MgO, affords a further illustration. Two electrons in the outer layer of magnesium are transferred to the oxygen layer of originally 6, thus bringing it up to 8. Counting the two electrons in the kernel, the total becomes equal to that of neon with an atomic number of 10; or averaging, in this case, the sum of the atomic numbers, 12 and 8, equals 10.

Sodium fluoride can be compared with magnesium oxide in the same way. The resulting ions of the former have their electrons arranged exactly like those of the latter, and A. W. Hull has verified this prediction by X-ray analysis of these two compounds, perfect crystalline identity being found. The relative stability of these compounds may be traced to the forces acting between the ions, and where double charges are involved, other things being equal, the stability may be greater as in the case of magnesium oxide, this compound having a high melting-point, low conductivity and solubility and greater hardness than that of sodium fluoride with which the comparison is made.

Sulphur affords a particularly interesting example in sulphur fluoride, $SF_6$, which is an odourless and tasteless gas with a low boiling-point of $-62°$ C. The inertness of this gas may be explained by the disposition of the fluorine ions round the sulphur ion in such a way as to prevent any appreciable stray field. It would be like a multipolar magnet with all the keepers in proper place, except that in the case of the atoms the attraction may not be magnetic in the sense of involving magnetic force as distinct from that known as electrostatic. It is to be noted that in this example the sulphur, which has 6 electrons in its outer layer, supplies 6 electrons to the 6 fluorine ions, thus completing 6 octets; and, as in sodium fluoride, the electrostatic fields are intense and the fluorine atoms sufficient in number to close completely round the sulphur ion and thus eliminate any stray field that would otherwise exist. Thus the conditions necessary for a gaseous state are fulfilled in this compound.

It will be seen that magnesium has a positive valence of 2, being this number of times in excess of that of neon (the Periodic Tables should, of course, be consulted whenever atomic values are given : such as atomic number, electrons in the outer part of the atom, etc.) which has an atomic number of 10. Similarly, sulphur has a positive valence of 6, having 6 more electrons than neon, but it has a negative valence of 2 because " it must take up more electrons before it can assume a form like that of the argon atom." The following rule may now be given :—

" The maximum positive valence represents the number of electrons which the atom possesses in excess of the number needed to form one of the particularly stable configurations of electrons. . . . The maximum negative valence is the number of electrons which the atom must take up in order to reach one of these stable configurations."

Before this theory can be made fairly complete, it is necessary to consider the principle of sharing electrons, as indicated in Chapter XIII, Figs. 17 and 18. The following statements taken from one of Langmuir's * papers will make this idea clear. " As a result of the sharing of electrons between octets, the number of octets that can be formed from a given number of electrons is increased. For example, two fluorine atoms, each having seven electrons in its outside shell, would not be able to form octets [in the joint sense] at all except by sharing electrons. By sharing a single pair of electrons, however, two octets holding a pair in common required only 14 electrons. This is clear if we consider two cubes with electrons at each of the eight corners. When the cubes are placed so that an edge of one is in contact with the edge of the other, a single pair of electrons at the ends of the common edge will take the place of four electrons in the original cubes [see Fig. 37]. For each pair of electrons held in common between two octets there is a decrease of two in the total number of electrons needed to form the octets.

" Let $e$ represent the number of electrons in the outside shell of the atoms that combine to form a molecule. Let $n$ be the number of octets that are formed from these $e$ electrons, and let $p$ be the number of pairs of electrons which the octets share with one another. Since every pair of electrons thus shared reduces by two the number of electrons required to form the molecules, it follows that—

$$e = 8n - 2p \qquad \text{. . . . . . . . . . . . (i)}$$

or

$$p = \tfrac{1}{2}(8n - e) \qquad \text{. . . . . . . . . . (ii)}$$

" This simple equation tells us in each case how many pairs of electrons or chemical bonds must exist in any given molecule *between the octets formed.* Hydrogen nuclei, however, may attach

* *Journ. of Industrial and Eng. Chem.* (April 1920).

themselves to pairs of electrons in the octets which are not already shared. For example, in the formation of hydrogen fluoride from a hydrogen atom and a fluorine atom there are 8 electrons in the shells ($e=8$). We place $n-1$ in the above equation and $p=0$. In other words, the fluorine atoms do not share electrons with each other. The hydrogen nucleus having given up its electron to the fluorine atom attaches itself to one of the pairs of electrons of the fluorine octet, and thus forms a molecule having a relatively weak external field of force. As a result, hydrogen fluoride is a liquid of low boiling-point instead of being salt-like in character.

" The equation given above is applicable to all types of compounds. For example, if we apply it to substances, such as sodium fluoride, sulphur fluoride, or potassium fluosilicate, which were previously considered, we find that in each case $p=0$. In other words, there are no pairs of electrons holding the atoms of these compounds together. On the other hand, if we consider the compound $N_2H_4$, we find that $p=1$. Since there are only two octets, the pair of electrons must be between the two nitrogen atoms while the hydrogen nuclei attach themselves to pairs of electrons of the nitrogen octet. It can be readily shown that this simple theory is, in fact, identical with the accepted valence theory of organic chemistry and leads to the same structural formulas as the ordinary theory in all those cases where we can take the valence of nitrogen to be 3, oxygen and sulphur 2, chlorine and hydrogen 1. In other cases, such as those where quinquivalent nitrogen has been assumed, the new theory gives results different from the old, but in each case in better agreement with the facts."

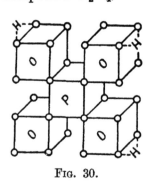

FIG. 30.

The rule of valence according to this theory may be stated thus :

1. Positive valence=the number of electrons an atom can give up.
2. Negative valence=the number of electrons an atom can take up.
3. Covalence=the number of pairs of electrons which an atom can *share* with its neighbours (=one to four atoms).

Corresponding examples are :—

1. Na, in sodium fluoride.
2. F, in sodium fluoride.
3. P, in orthophosphoric acid, $H_3PO_4$, in which $n=5$, $e=32$, $p=4$. See Fig. 30.

The above phosphorous compound, though a little complicated (for simpler examples see Figs. 32 and 33), illustrates both 'sharing' and 'complete giving' of electrons, as already explained. There are 3 hydrogen atoms which give up their electrons to 3 oxygen atoms and thereby bring the total number of electrons in the outer parts of the oxygen atoms up to 7. Since phosphorus has 5 outer electrons, it becomes now possible to construct the molecule, as illustrated, without leaving an empty corner, and thus all the octets of P and O are completed.

The above compound may be represented by the constitutional formula :—

In a similar manner the following compounds (acids) may be studied :—

| | |
|---|---|
| Hypophosphorous | $H_3PO_2$ |
| Phosphorous | $H_3PO_3$ |
| Pyrophosphorous | $H_4P_2O_5$ |
| Hypophosphoric | $H_4P_2O_6$ |
| Metaphosphoric | $HPO_3$ |
| Orthophosphoric | $H_3PO_4$ |
| | (as above) |
| Pyrophosphoric | $H_4P_2O_7$. |

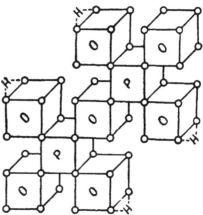

<small>FIG. 31.</small>

The last acid has an octet structure, as shown by Fig. 31.

S. C. Bradford,* in giving a short account of this theory, describes the water molecule in the following words : " The use of formula (ii) will be clear if we employ it to determine the structure of a few compounds. The internal arrangement of water molecules is of great interest from its wide application as a solvent. Since the two hydrogen nuclei always tend to hold pairs of electrons and never octets, we may take $n=1$ for the oxygen atom. Then there are 6 available electrons in the oxygen atom and 2 in the hydrogen atoms, making $e=8$, whence equation (ii) gives $p=0$. This means that no electrons are held in common between octets, which is obvious in this case, since there is only one octet. The two hydrogen nuclei are held by electrostatic attraction to two pairs of the electrons forming the octet. The water molecule may, therefore, be pictured as a cube with two hydrogen nuclei hanging on to opposite edges. This structure indicates that water forms mole-

* *Science Progress* (1920), 15, p. 50.

cules in which the electrostatic forces are almost completely compensated internally. All the electrons form an octet, and hence the molecule should have a rather weak external field of force. Water should, therefore, be easily volatile and should not be a good conductor of electricity. But the less symmetry of the molecule, as compared with the neon atom, shows that the boiling-point should be much higher."

In considering the nitrogen compounds, $NH_3$ is in accordance with this theory, and $N_2H_4$ may be represented thus—

$$H_2N-NH_2,$$

while NH, $NH_2$, $NH_4$ and $NH_5$ cannot exist. This will be evident when applying the formula (ii) to such hypothetical structures, since it would indicate impossible values for $p$. The oxides of nitrogen, so well known, are shown to possess properties in full

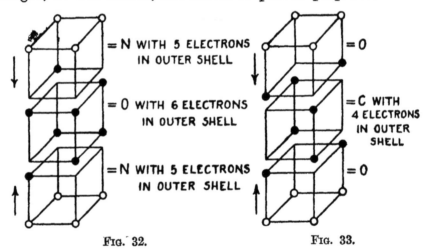

= N WITH 5 ELECTRONS IN OUTER SHELL

= O WITH 6 ELECTRONS IN OUTER SHELL

= N WITH 5 ELECTRONS IN OUTER SHELL

= O

= C WITH 4 ELECTRONS IN OUTER SHELL

= O

FIG. 32.          FIG. 33.

accord with this theory, and, moreover, compounds which had not been associated with one another, on account of their properties being similar, are brought into parallel relationship by means of this theory. Two interesting cases are $N_2O$ and $CO_2$. In Figs. 32 and 33 each atom is provided with the number of outer electrons indicated by this theory (see Table XI or XII), the black discs or spheres representing those electrons which share equally a position in both octets when the atoms come together as indicated by the arrows. The stability and similarity of these two molecules is at once apparent and, on comparing their various properties, it will be seen how remarkably alike they are (see Table XIII).

In addition to the following data there are similarities in the hydrates formed, viz. $N_2O \cdot 6H_2O$ and $CO_2 \cdot 6H_2O$. The vapour pressure of the first hydrate is 5 atmospheres at 6° C., while the second hydrate has this pressure at —9° C. The respective heats of formation are 14,900 and 15,000 calories per gram-molecule. The surface tensions per centimetre are the same at 12·2° C. and

9° C. respectively, the value being 2·9 dynes. There is a freezing-point difference, but this is to be expected even when there are only slight differences in structure.

TABLE XIII

| Property | $N_2O$ | $CO_2$ |
|---|---|---|
| Critical pressure . . . . . . | 75· | 77· |
| Critical temperature deg. C. . . . . | 35·4 | 31·9 |
| Viscosity at 20° C. . . . . . . | $148 \times 10^{-6}$ | $148 \times 10^{-6}$ |
| Heat conductivity at 100° C. . . . . | 0·0506 | 0·0506 |
| Density of liquid at − 20° C. . . . . | 0·996 | 1·031 |
| Density of liquid at +10° C. . . . . | 0·856 | 0·856 |
| Refractive index of liquid (radiation giving D line used) . . . . . . | 1·193 | 1·190 |
| Dielectric constant of liquid at 0° C. . . | 1·598 | 1·582 |
| Magnetic susceptibility of gas at 40 atmos. 16° C. . . . . . . . | $0·12 \times 10^{-6}$ | $0·12 \times 10^{-6}$ |
| Solubility in water 0° C. . . . . | 1·305 | 1·780 |
| Solubility in alcohol 15° C. . . . . | 3·25 | 3·13 |
| Molecular number * . . . . . | 22· | 22· |

Carbon monoxide (CO) and nitrogen ($N_2$) exhibit remarkable similarity, as shown by Table XIV.

TABLE XIV

| Property | CO | $N_2$ |
|---|---|---|
| Freezing-point in degrees absolute . . . | 66· | 63· |
| Boiling-point in degrees absolute . . . | 83· | 78· |
| Critical temperature in degrees absolute. . | 122· | 127· |
| Critical pressure in atmospheres . . . | 35· | 33· |
| Critical volume . . . . . . | 5·05 | 5·17 |
| Solubility in water at 0° C. . . . . | 3·5 | 2·46 |
| Density at boiling-point . . . . . | 0·793 | 0·797 |
| Viscosity at 0° C. . . . . . . | $163 \times 10^{-6}$ | $166 \times 10^{-6}$ |
| Molecular number . . . . . . | 14· | 14· |

The great stability of $N_2$ which resembles argon, the freezing-points being $N_2 = 63°$, $Ar = 85°$ (degrees centigrade on the absolute scale) while the boiling-points on the same scale are nearly the same, seems to suggest that the nitrogen molecule is surrounded by a single octet like the outer shell of argon. These two molecules (CO and $N_2$), which are very stable, are shown by Figs. 34 and 35. It is to be noted that the CO-molecule, as shown, would readily accommodate another oxygen atom at the top of C, thus forming $CO_2$ as shown above. Similarly, NO is very stable, its molecule being as shown by Fig. 36. In this figure the upper front electron may be wrongly placed, for it may occupy the space below, leaving an open corner on the top.

* Molecular number = sum of atomic numbers.

It occurs to the writer that a polar electron from each nitrogen atom might be drawn to the corners (junctions) A and B in Fig. 35, and thereby increase the stability of the molecule. No doubt Langmuir has considered such a possibility and, as no specific mention has been made of such a circumstance, there are no doubt reasons for not entertaining this idea. Indeed Langmuir says, in discussing the first long period in the Periodic Table, that there is no necessity for, in fact every probability against, the kernel being concerned in the process of transferring electrons from one part of the kernel to another or to the outside shell, or even a transfer taking place between two separate outer shells. Lewis had suggested that the kernel might not be uniquely and permanently defined and that the transfer of electrons as just indicated, whereby electrons become suspended midway between two such stages, might become responsible for the absorption of light, as in coloured salts. Langmuir attributes the absorption of light to the transfer of electrons between different parts of the same outside shell, if

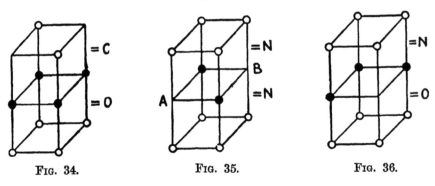

Fig. 34.             Fig. 35.             Fig. 36.

not to the ease with which an electron is gained or lost by the outside shell.

Substances showing like properties, as illustrated above (see Tables XIII and XIV), Langmuir calls *isosteres*, or *isosteric* substances.

The oxides of nitrogen are of special interest, and this theory indicates that the following structures are most probable :—

$N_2O_2$ = $O{=}N{-}N{=}O$
$n=4$    $e=22$      $p=5$

$N_2O_3$ = $O{=}N{-}O{-}N{=}O$ or $\underset{O}{\overset{O}{>}}N{-}N{-}O$

$n=5$    $e=28$      $p=6$

$N_2O_4$ = $\underset{O}{\overset{O}{>}}N{-}N\underset{O}{\overset{O}{<}}$

$n=6$    $e=34$      $p=7$

$N_2O_5$ = $\underset{O}{\overset{O}{>}}N{-}O{-}N\underset{O}{\overset{O}{<}}$

$n=7$    $e=40$      $p=8$

$N_2O_6$ does not seem probable.

NOTE. — The precise positions of the atoms may differ somewhat from these arrangements. These formulæ will be appreciated when drawing the octets and fitting the atoms together.

It will be seen that symmetry plays an important part in this theory, and it is interesting to note that the activity of the fluorine molecule may be due in part to the ease with which the atoms can be separated, as in many cases the stronger chemical reactivity is traceable to the atom as distinct from the molecule unless it is in the ionic state. The fluorine molecule may be pictured in the act of formation by Fig. 37.*

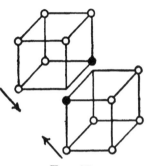

FIG. 37.

When the atoms are joined the two electrons nearest together, shown black, share places in the two octets, thus completing them; but this particular combination from a mechanical point of view is less stable than some of those shown above, and a comparatively small force would separate the atoms. Now oxygen, which is less active than fluorine, has a more stable structure mechanically, as shown by Fig. 38.

It should be realised that these figures are largely diagrammatic and the inferences to be drawn from them as regards stability, etc., may break down if pressed too far. They represent the nearest approximation to the truth which can at present be made. When more is known about the atomic and molecular architecture the specification will probably be drawn up much on the same lines but with a greater wealth of detail. Moreover, one has to consider atomic combinations from the engineer's point of view, as if they were complex steel structures subject at times to excessive vibration, like, for example, a vibrating mechanism, and to imagine certain configurations favouring the culmination of vibrations at certain points. This means that one structural configuration will be more stable than another. Natural periods of vibration of certain parts is suggested which would determine stability—to a certain extent at any rate.

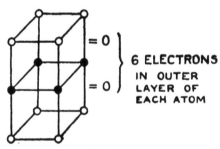

= O
= O
} 6 ELECTRONS
IN OUTER
LAYER OF
EACH ATOM

FIG. 38.

* It is to be noted that in most figures shown, no attempt is made to represent the kernel or inner parts of the atom, as it is only the outer electrons which take part in chemical action.

# CHAPTER XV

## THE OCTET THEORY: BEING A CONTINUATION FROM THE PRECEDING CHAPTER: IONISATION IN SOLIDS AND LIQUIDS: ETC.

The octet theory involves so many considerations that it was thought desirable to group a number of them together, with sub-headings for convenience of reference. This has the advantage of not unduly overloading the previous treatment with details which are apt to confuse the main points discussed.

**Ionisation in Solids and Liquids.**—It has long been held that ionisation takes place only when a solvent is used to separate the atoms or molecules from one another whereby they become ions. In view of the work of S. R. Milner,[*] J. C. Ghosh [†] and N. Bjerrum,[‡] together with Bragg's measurements of the spacing of atoms in crystals, it seems necessary to extend the principle of ionisation to solids and strong electrolytes in which the $+$ and $-$ ions represent charged atoms or molecules existing as such in all physical states which retain the structure similar to that found in certain crystals. This principle is in conformity with the octet theory, as will be seen from the statements in the main paragraphs following Table XII on page 102.

Referring to the recent work of Ghosh in particular, this leads to a modification of the well-known assumption of Arrhenius: that undissociated molecules and ions were present in the case of strong electrolytes, the latter being few in number. X-ray measurements of the crystal lattices of salts, such as KCl, NaCl, LiCl, etc., have been the means of providing precise values for the distances between oppositely charged atoms or ions. Making use of these data and assuming that the marshalling of the ions in an aqueous solution is analogous to that obtaining in crystals—or obtaining when the solution itself is crystallised—Ghosh has devised formulæ which give support to the theory that ions are held together as ions by a strong electrostatic force of attraction. Upon applying Maxwell's law of distribution of velocities, it was shown that only

---

[*] *Phil. Mag.* (1918), 35, p. 214.
[†] *Chem. Soc. Journ. Trans.* (1918), 113, p. 449.
[‡] *Zeit. Electrochem.* (1918), 24, p. 321.

those ions will be free which escape from the electrostatic binding fields, these escaping ions being endowed with sufficient kinetic energy (energy of motion) to free themselves ; and these free ions become the means of effecting electrical conduction through the electrolyte. The number of such free ions is deduced from equations involving Avogadro's number and the dielectric constant of water, etc. It is assumed that the ions of a salt-molecule form a completely saturated doublet, and that work is performed in separating them ; and at infinite dilution the ions are sufficiently separated to exercise no electrostatic binding influence on one another.

**The Structure of H₂.**—On the view that a revolving electron is a negative charge of electricity moving in a closed circle, a magnetic effect should result since this action is equivalent to a circular current which is known to produce a magnetic field. Accepting Bohr's theory of the hydrogen atom, in constructing the hydrogen

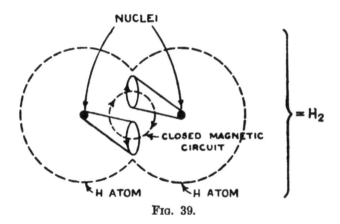

Fig. 39.

molecule, the movement of the electrons in circles between or round the hydrogen nuclei would give the molecule paramagnetic properties (unless the movement is such that each electron sets up a mutually opposing magnetic field), whereas hydrogen is diamagnetic. To meet this difficulty A. E. Oxley * in discussing the Lewis-Langmuir theory, which can be brought into harmony with Bohr's by assuming that the electrons revolve round the stable positions † in the atom, has suggested that the electrons revolve in small circles, as shown in Fig. 39. It will be seen that in this case the smallness of the orbit is in conformity with what one might expect from the octet theory which would seem to involve electrons revolving round ' corner positions,' but if they revolve in opposite directions, as here shown, a small closed magnetic circuit would be produced, as indicated by the dotted circle with arrow heads. Thus there should be no external polarity to the molecule as a whole, while the

* *Nature* (1920), 105, p. 327 ; see Oxley, *Roy. Soc. Proc.* (1920), 98, p. 264.
† *i.e.* round centres in cubic symmetry if extended to other elements

magnetic field would tend to bind the nuclei together, for the electrons would be attached to their original nuclei (this force is represented by the converging lines), and the magnetic field (small dotted circle) would link the electrons themselves together. Such a molecule might be diamagnetic and the scheme referred to as a *diamagnetic coupling*. A *paramagnetic coupling*, by way of illustration, is shown by Fig. 39A. This coupling may not, of course, be a possible one for the hydrogen nuclei, but it will serve to show a possible type for other atoms. These terms are those used by Oxley in the papers referred to on the previous page.

**First Long Period in the Periodic Table.**—In the first short period, Li to F, Langmuir says : Considering C, N and O, " the properties of the atoms up to this point [those of H, Li, Be and B] have been determined by their ability to give up one or more electrons. With carbon and the elements which follow it there is less tendency to part with electrons, and more tendency to take up electrons to form a new octet. This opens up new possibilities in the formation of compounds and as a result we find a remarkable contrast between the properties of oxygen and nitrogen and those

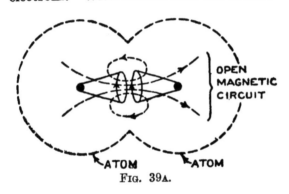

OPEN MAGNETIC CIRCUIT

ATOM    ATOM

FIG. 39A.

of lithium and beryllium. Among compounds of carbon with hydrogen and oxygen the valence almost without exception can be taken as four for carbon, two for oxygen and one for hydrogen. . . . When nitrogen is introduced into organic compounds there is much uncertainty in using this theory of valence [the ordinary one]."

Very much as there are ' turning-points ' at the inert gases there are ' places ' in the approximate centres of each period, and as the periods lengthen, owing to more cells existing in the atom, these places become longer and thus include more elements. The analogy (taken in converse sense) is imperfect, but the following Figs. 40, 41 and 42 will make this idea clearer.

Beyond titanium there is a region where most periodic relations fail and, as bromine is reached, the normal properties return. While the above figures do not fully illustrate this statement, they show that the middle elements of each period typify such a ' change,' and it is to be expected that the absence of certain characteristics would be shared by those elements which are near to the central ones. This is to be expected from the octet theory, because of the increase in cells in this series, the middle members not tending so strongly to revert to the perfectly inert type either by giving up or by taking up electrons.   Iron is in one sense an irregular member since it has a complete octet (see Tables XI and XII), but though it has room for more electrons in its outer layer, it has a weak tendency to take up electrons and also little tendency to part with any, while the elements Ti, V, Cr and Mn are predominantly electro-positive in character.   A certain degree of inertness is manifest in iron though the element with which it is associated must be con-

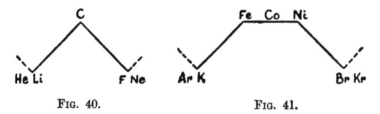

FIG. 40.                              FIG. 41.

sidered, as some iron compounds are comparable to those composed of the more active elements.

In Table XI the difference between the properties of these elements and those of the corresponding elements of the earlier periods is indicated by the heavy lines enclosing them.   It should be noted that for elements having a considerable number of external electrons, say 5, 6 and 7 as in V, Cr and Mn, they would have to part with these to revert to the argon type, but such a tendency is not a strong one with what might be termed the middle elements in the long periods (K, Ca and Sc form a contrast to these in easily

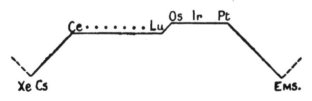

FIG. 42.

parting with their electrons and thus reverting to the most positive ionic type).   As a matter of fact, those under consideration only form bivalent ions and tervalent cations, while they do not form univalent or quadrivalent ions in solution.

The lack of definite forces to determine the distribution of the electrons amongst the cells renders these elements capable of ab-sorbing light (Lewis) in the visible spectrum.   See in this connection A. W. Stewart's Periodic Table given in his book, *Recent Advances in Physical and Inorganic Chemistry* (1920), where a table is shown bringing into prominence the 'middle elements' which form coloured salts.

In cobalt and nickel with atomic numbers respectively one and two units more than in iron, the extra electrons cannot arrange themselves over those of the underlying shell. In nickel there are 5 electrons in each hemisphere and these tend to take up positions exactly over the 4 underlying electrons. The extra electrons go to the poles (one to each) and thereby a symmetrical arrangement is realised. In cobalt there is one electron only which goes to *one* pole. Thus it will be seen that although these elements have 10 and 9 electrons in the outer layer of the atom they cannot exercise valences corresponding to these numbers, and the present theory explains satisfactorily their anomalous position in the Periodic Table.

Langmuir suggests that the introduction of the electrons in the polar axis tends to force the outer electrons of the zones away from their position over the underlying ones, and this weakens the magnetic forces of the atom, the 4 electrons probably suffering an orientation of 45° about the polar axis. This arrangement of electrons gives effect to the β-form (see Table XI) which has a higher degree of symmetry than the α-form in which the 4 electrons were closer to the inner ones. Langmuir says : " According to postulate 3 we should look upon the transformation from the α-form to the β-form as involving the passage of electrons between different cells in the outer layer. It is perhaps best to imagine that it is the empty cells which arrange themselves over the underlying electrons in the β-form."

Copper, zinc and gallium, or those elements leading up to selenium, tend to give up electrons in such a way as to revert to the symmetrical β-form of nickel. The β-form is more stable as it resembles more closely nickel, and on this account the β-atoms are placed in a line with the inert gases in Table XI, but they still have a strong residual field as their outer layer, though containing a stable group of electrons, has room for more.

In arsenic, selenium and bromine the electronegative character strongly asserts itself owing to their closeness to krypton ; and thus they resemble phosphorus, sulphur and chlorine ; but their tendency to form insoluble secondary valence products distinguishes them from those of the short period.

**Second Long Period.**—Beyond krypton the second layer forms, which layer is not fully completed until xenon is reached, the elements involved in this series forming the second long period. Referring to postulates 3, 4 and 5, and the characteristic features of the first long period, it will be seen that the second period is very similar ; but the symmetry is less perfect in the second series, and there is a tendency to form insoluble salts and secondary valence compounds as in the first period.

Considering palladium, probably its two extra electrons over that of an octet go to the poles. In the elements beyond palladium, " the electrons around the polar axis seek positions as far as possible from the electrons in the second shell, so that the atoms tend to revert to the $\beta$-form of the palladium atom. Thus silver forms colourless univalent ions, cadmium bivalent ions, etc. These properties and their explanation are so nearly like those of the first long period that we need not consider them in more detail."

**The Rare-Earth Period.**—This series begins after xenon and terminates in the chemically stable emanations. The first 3 or 4 elements have a strong electropositive character, as would be expected. As more electrons are added these arrange themselves over the 18 underlying electrons of the third shell in a manner similar to those of the first long period which take up positions over the 8 underlying ones of the second shell. This series yields a set of elements similar to Ti, V, Cr, Mn, Fe, Co and Ni. By referring to Table XI it will be seen that this family corresponds to those of the rare-earths. The 18th element from xenon is lutecium and this element marks definitely the last of the rare-earth elements.

It is highly important to follow closely any suggestions which will help to elucidate the nature of the rare-earth group under consideration, as it represents one of the chief outstanding problems in chemistry. One cannot do better therefore than to quote Langmuir: "Since the forces holding these 18 electrons are predominantly magnetic and since the constraints are not of the rigid kind characteristic of the inert gases, we should expect these elements to be paramagnetic. As a matter of fact, the rare earths are the most strongly paramagnetic of any of the elements except those from manganese to nickel. Even barium begins to show a perceptible paramagnetism (strontium is diamagnetic). The susceptibilities of only a few of these metals have been determined [in the metallic state], but the atomic susceptibilities of cerium, praseodymium, neodymium and erbium are, respectively, 2, 7, 11 and 7 times that of manganese. Gadolinium sulphate lies between ferric sulphate and manganese chloride in magnetic properties.

" It seems probable that the most marked magnetic properties occur with the elements samarium, europium and gadolinium, for these are the eighth, ninth and tenth elements from xenon and thus should correspond most closely in their structure to iron, nickel and cobalt. In samarium there is probably a slight tendency for the eight electrons in the outside layer to arrange themselves at the corners of a cube, while in gadolinium the two extra electrons are in the polar axis. But in other properties these 3 elements should not differ radically from the other rare earths.

" By the time the 18 electrons have been added the electrostatic forces have begun to oppose the magnetic attraction to a marked degree. Therefore, when in tantalum an additional electron is added, the whole outside shell tends to rearrange itself so that the empty cells will come opposite the electrons of the underlying shell. The most symmetrical arrangement of this kind will occur when there are 18 empty cells opposite the 18 underlying electrons. The atomic number of niton [radium emanation] in which the fourth shell is complete is 86—therefore an element having 18 empty spaces in the fourth will have an atomic number of 68 corresponding to erbium. The structure of this β-form of erbium has the same kind of stability for large nuclear charges that we found in the case of β-nickel and β-palladium. We may therefore expect that the atoms beyond lutecium will show a marked tendency to revert to β-erbium. Thus tantalum with an atomic number 73 tends to lose 5 electrons and tungsten to lose 6. The properties of tantalum and tungsten thus resemble those of columbium [niobium] and molybdenum, but because of the complexity of the atom to which they revert, and in general because of the large numbers of electrons in their outside shell, their secondary valence forces are more highly developed.

"In accordance with the marked change in the electron arrangement beyond lutecium we find that the paramagnetism is practically absent in the elements tantalum and tungsten.

"The β-form of the erbium atom contains 18 empty cells arranged over the 18 cells of the third cell. When electrons are added as we pass to elements of large atomic number the first 8 of them naturally tend to arrange themselves at the corners of a cube, because of the magnetic attraction of the 8 electrons in the *second* shell. The next two electrons for reasons of symmetry then arrange themselves in the polar axis. We thus have the 3 'eight-group' elements osmium, iridium and platinum. Because of the weakness of the forces acting between the fourth and the second shells we should not expect strongly developed magnetic properties in these elements. As a matter of fact, osmium and iridium have susceptibilities nearly equal to zero, but there is a small but sharp rise at platinum, making this element about ¼ as paramagnetic as palladium. The next elements, gold, mercury, etc., are distinctly diamagnetic. The same sharp break occurs here as we found between nickel and copper, palladium and silver, lutecium and tantalum, although its magnitude is much less. We therefore assume that beyond platinum the electrons tend to rearrange themselves in a β-form in which the 10 electrons that have been added since erbium endeavour to get farther away from those of the underlying electrons. The 8 empty cells tend to take symmetrical positions in the atom probably at the corners of a cube, and the cells containing electrons space themselves as best they can. The fact that an arrangement of this kind does not have nearly the symmetry which we found for the β-form of the nickel atom is probably the explanation of the fact that the tendency of the succeeding elements to revert to this β-form of platinum is much less marked than we observed in the case of reversion to nickel, paladium and erbium. Thus we find that gold and mercury have variable valence differing in this respect from silver and cadmium. Thallium forms univalent and tervalent ions, whereas indium forms only tervalent. Lead only exceptionally is quadrivalent, while this seems to be the normal condition of tin compounds. Thus stannous salts are strong reducing agents, but bivalent lead salts are not. Bismuth is normally tervalent and forms only a few very unstable compounds in which it is quinquivalent. Antimony, on the other hand, has about equal tendencies to be tervalent or quinquivalent.

"There is an interesting sudden break in the susceptibility curve between lead and bismuth. Gold, mercury and thallium are very slightly diamagnetic, but bismuth is the most strongly diamagnetic element with the exception of the inert gases. In all the elements between gold and niton the positions of the electrons are determined mainly by electrostatic forces (postulate 6). But magnetic forces still tend to cause the electrons to arrange themselves in the 8 available cells (at platinum) so that they will be placed as symmetrically as possible with respect to the underlying electrons. Now the 4 additional electrons (in lead) can arrange themselves in the 8 spaces with reasonable symmetry, but the 5 electrons in bismuth cannot do so. The extra electron displaces the others and thus weakens the magnetic forces and strengthens the electrostatic. In agreement with this theory we find that there is a similar, although smaller, minimum in susceptibility at phosphorus, arsenic and antimony, the elements which also have atomic numbers 3 less than those of the following inert gases. We also find distinct maxima at germanium, tin and lead, which have 4 electrons less than the inert gases which follow them."

**A General Observation.**—In conformity with this theory it is clear that, as the elements approach those of the inert group from either side, they are more nearly alike, as instanced in Na, K, etc., on one side, and Cl, Br, etc., on the other. This being the case, one should not be surprised to find the elements farthest removed from those of the stable type, to which they tend to revert, somewhat exceptional in character; these elements being modified by new conditions which arise from the progressive expansion of the atom, as indicated above. One can imagine a point being reached in which the 'middle' elements are so far removed from those at the 'turning-points' (see Figs. 40, 41 and 42) as to fall into a sort of 'hesitating' state giving rise to a state of equality, as instanced by those of the rare earths.

Why such elements do not appear in the last series containing the radio-active elements is a question one might ask. The considerable number of leads as a result of disintegration does not afford a proper answer to this question, but it is perhaps suggestive in this connection.

**Valence, Co-ordination Number and Covalence.**—The octet theory indicates that each carbon atom of a molecule of an organic substance shares all 4 pairs of electrons with adjacent atoms, thus completing its octet. For such substances a pair of electrons held in common corresponds to the bond in the ordinary valence theory. Among other compounds this relationship does not hold. This theory indicates that in nitric acid the nitrogen shares 4 pairs of electrons with the oxygen atoms, so that the valence of nitrogen is 4 ; but to distinguish between the valence thus indicated, in the case of nitrogen, and that of ordinary valence as established prior to this theory, the term *covalence* will be used to denote the pairs of electrons which a given atom shares with its neighbours.

Now Werner's *co-ordination number* represents the number of atoms or molecules, irrespective of their valence, which arrange themselves round a given atom ; and the maximum co-ordination number for carbon is 4, which is realised in many compounds as, for example, saturated hydrocarbons and halogen compounds. There are substances, however, in which the co-ordination number is less than 4, e.g. $CO_2$, formaldehyde, etc.

The octet theory does not assign definite valences to atoms, except in the case of the maximum positive valence, representing the number of outer electrons in the shell, which is a definite conception, as is likewise the maximum negative valence representing the number of electrons an atom may take up to become stable like the inert gases. In most compounds, however, the atoms do not take up or give up electrons, but share them in the manner already indicated. Thus there are *three* principles involved in this theory.

With carbon it so happens that the number of pairs of electrons shared by other atoms is equal to both the maximum positive and maximum negative valence. With other elements there is no necessary relation between the number of pairs of electrons shared and the number of electrons in the shell of the original atom ; and it is for this reason that confusion occurs when the ordinary valences are applied to *inorganic* compounds.

As already shown, Langmuir's equation affords a means of determining from the total number of available electrons the pairs of electrons held in common ($p$), but the theory does not explain why certain compounds exist to the exclusion of others—that is to say, there are cases which require an extension to the theory. For example, phosphorus and nitrogen contain the same number of available electrons in their shells, yet the compounds of these two elements do not run parallel, for there are no compounds answering to $H_3NO_4$, $Na_4N_2O_7$ and $P_2O$, as might be expected.

It is therefore necessary to introduce covalence values as

observed for different elements and thereby extend the usefulness of the theory.   To this end Table XV is prepared, in which—

$B$=available electrons in the shell of the atom (see Table XI or XII).

$P$=the number of *pairs* of electrons which the atom shares with other atoms.

$S$=the maximum number of pairs of electrons which an atom can share with one other atom.

$0+$=inability of the atom to share electrons with other atoms; but it gives up one or more electrons and becomes positively charged, as in the case of the lithium ion described above.

$0-$=ability of the atom to take up electrons to complete its octet; but the atom does not share electrons with other types of atoms.   The Cl$^-$ ion is an example.

### TABLE XV

#### COVALENCE OF THE FIRST 18 ELEMENTS

| Element | $B$ | $P$ | | | | | | $S$ |
|---|---|---|---|---|---|---|---|---|
| H | 1 | 0+ | 1 | | | | 0- | |
| He | 0 | 0 | | | | | | |
| Li | 1 | 0+ | | | | | | |
| Be | 2 | 0+ | 4? | | | | | |
| B | 3 | 0+ | 4 | | | | | |
| C | 4 | | 4 | | | | | 3 |
| N | 5 | | 4 | 3 | 2 | | | 3 |
| O | 6 | | | 3 | 2 | 1 | | 2 |
| F | 7 | | | | | 1 | 0- | 1 |
| Ne | 0 | 0 | | | | | | |
| Na | 1 | 0+ | | | | | | |
| Mg | 2 | 0+ | | | | | | |
| Al | 3 | 0+ | 4 | | | | | |
| Si | 4 | 0+ | 4 | | | | | 2 |
| P | 5 | 0+ | 4 | 3 | | | | 2 |
| S | 6 | 0+ | 4 | 3 | 2 | 1 | 0- | 2 |
| Cl | 7 | | 4 | 3 | 2 | 1 | 0- | 1 |
| Ar | 0 | 0 | | | | | | |

The covalence of those elements from carbon to neon decreases as the number of electrons approaches that of neon, while the corresponding elements of the next period retain a maximum covalence of 4.   This is in accordance with the fact that $HClO_4$ is formed whilst no corresponding compound with fluorine exists.

There is a close relationship between covalence and Werner's co-ordination number, this number being 4 in the compounds : $NH_4Cl$, $HBF_4$, $H_2SO_4$, $H_3PO_4$, $HPH_2O_2$, $H_2PHO_3$ and $HClO_4$, but not four in $NHO_3$, $CO_2$ and $CH_2O$.   In the octet theory the covalence of the central atom of all these compounds is 4.   " In $HNO_3$, $CO_2$ and $CH_2O$ one or more oxygen atoms is held to the central atom by two pairs of electrons, while in all the others there is never more than one pair of electrons involved in holding together two adjacent atoms"; and the difference between the two theories accounts for many of the cases of unsaturated supplementary valences.

Langmuir says: "In a very great number of compounds the co-ordination number and the covalence are identical, and the octet theory then corresponds exactly to Werner's theory, just as for organic compounds it is equivalent to the ordinary valence theory."

A few examples will make the foregoing statements clearer. In $HBF_4$ $n=5$, $e=32$, $p=4$. Each F-atom thus shares a pair of electrons with the octet of the B-atom. The covalence of B is 4 in this compound. In $NH_4Cl$ $n=2$, $e=16$, $p=0$. The 4 H-nuclei thus attach themselves to the 4 pairs of electrons in the N-octet, giving rise to the positive ion $NH_4^+$; and since the Cl-atom has a completed octet it becomes a negative ion $Cl^-$. Therefore ammonium chloride is a salt resembling sodium chloride. It will be seen that the central nitrogen atom has a covalence of 4. These structures correspond exactly to those given by Werner.

$B(CH_3)_3NH_3$ is of interest as its constitution is not given by the ordinary valence theory. In this case $n=5$, $e=32$, $p=4$. The structure is, therefore,

$$\begin{array}{ccc} H_3C & & NH_3 \\ & \diagup B \diagdown & \\ H_3C & & CH_3 \end{array}$$

This is a typical primary valence compound in no way different from organic compounds. Carbon and nitrogen here have the same covalence value 4. In Werner's theory the bond between the B-atom and N corresponds to a *supplementary* valence, the others being of the primary type. According to the octet theory they are all of the same type.

Of further interest is the compound $B(CH_3)_3$ in which $n=3$ (not 4, because it would be impossible to hold 3 methyl groups by four pairs of electrons), $e=24$ and $p=0$. The structure, therefore, becomes (with $n=3$) $B^{+++}$ $[CH_3]_3^-$. The volume of the B-atom being small, as compared with the methyl groups which surround it, the external field of this compound is not strong and it has, therefore, a low boiling-point and does not become an electrolyte.

In the platino-ammonia compounds Pt is bivalent in that there are two available electrons in the outer shell of the atom. In $Pt(NH_3)_4Cl_2$, $n=7$, $e=48$, $p=4$. Four $NH_3$ radicles are thus attached directly to the Pt, each sharing a pair of electrons. In this compound the N and the Pt atoms have a covalence of 4 and the Cl atoms become ions; and this one, as well as other similar compounds, affords an example of primary valence compounds according to the octet theory.

**Ions of the Sulphur Acids.**—The constitution of oxyacids are of interest. The ions are here given, as many of the acids do not exist in the free state. In determining $e$ the charge on the ion must be taken into account. " In the more stable acids all 4 pairs of electrons in the octets of the central atoms are shared by adjoining atoms." The connecting lines do not, of course, represent ordinary bonds : they represent pairs of electrons shared ($p$).

Sulphurous $SO_3^{--}$        $n=4$   $e=26$   $p=3$

$$O-S{\Large\langle}{}^{O}_{O}$$

Sulphuric $SO_4^{--}$        $n=5$   $e=32$   $p=4$

$$\begin{array}{cc} O & O \\ & \diagdown\!\!\diagup \\ & S \\ & \diagup\!\!\diagdown \\ O & O \end{array}$$

Hyposulphurous $S_2O_4^{--}$   $n=6$   $e=38$   $p=5$

$$\begin{array}{cc} O & O \\ \diagdown\!\! & \!\!\diagup \\ S\!\!-\!\!S & \\ \diagup\!\! & \!\!\diagdown \\ O & O \end{array}$$

Thiosulphuric $S_2O_3^{--}$       $n=5$   $e=32$   $p=4$

$$\begin{array}{cc} O & S \\ & \diagdown\!\!\diagup \\ & S \\ & \diagup\!\!\diagdown \\ O & O \end{array}$$

Pyrosulphuric $S_2O_7^{--}$      $n=9$   $e=56$   $p=8$

$$\begin{array}{ccc} O & & O \\ | & & | \\ O-S & -O- & S-O \\ | & & | \\ O & & O \end{array}$$

Dithionic $S_2O_6^{--}$        $n=8$   $e=50$   $p=7$

$$\begin{array}{cc} O & O \\ | & | \\ O-S & -S-O \\ | & | \\ O & O \end{array}$$

Trithionic $S_3O_6^{--}$        $n=9$   $e=56$   $p=8$

$$\begin{array}{ccc} O & & O \\ | & & | \\ O-S & -S- & S-O \\ | & & | \\ O & & O \end{array}$$

Tetrathionic $S_4O_6^{--}$    $n=10$   $e=62$   $p=9$

$$O-\overset{\overset{\displaystyle O}{|}}{\underset{\underset{\displaystyle O}{|}}{S}}-S-S-\overset{\overset{\displaystyle O}{|}}{\underset{\underset{\displaystyle O}{|}}{S}}-O$$

Pentathionic $S_5O_6^{--}$    $n=11$   $e=68$   $p=10$

$$O-\overset{\overset{\displaystyle O}{|}}{\underset{\underset{\displaystyle O}{|}}{S}}-S-S-S-\overset{\overset{\displaystyle O}{|}}{\underset{\underset{\displaystyle O}{|}}{S}}-O$$

Persulphuric $S_2O_8^{--}$    $n=10$   $e=62$   $p=9$

$$O-\overset{\overset{\displaystyle O}{|}}{\underset{\underset{\displaystyle O}{|}}{S}}-O-\overset{\overset{\displaystyle O}{|}}{\underset{\underset{\displaystyle O}{|}}{S}}-O$$

The persulphuric ion may be illustrated by Fig. 43, which shows the 'octets' sufficiently separated to indicate the parts of the atoms about to share electrons.

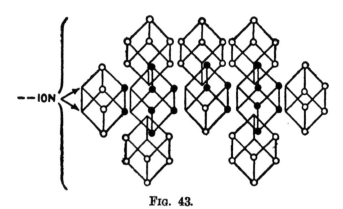

--ION

FIG. 43.

**Silicon.**—The $SiF_6^{--}$ ion can exist in solution or it can take up two hydrogen ions forming $H_2SiF_6$. In the case of complex silicates the conception of a *secondary valence* arises, since to account for $SiF_6^{--}$ it is necessary to assume that the fluorine atoms can "fit opposite the 6 faces of the cube" which involves "purely geometrical and electrical factors." Langmuir says : "The fluosilicate ion has a structure exactly like that of the sulphur fluoride molecule, since the number and arrangement of the electrons are the same. This idea is clear if we consider that the atomic number of silicon is 14, while that of sulphur is 16. Thus, if we should replace the nucleus of a silicon atom, without disturbing any of the surrounding electrons, we would have removed two positive charges and we would obtain a negative ion with two negative charges of the formula given above. In the presence of potassium ions we would then have the familiar salt potassium fluosilicate. The theory is thus capable of explaining typical complex salts. In fact, it is applicable to the whole field of inorganic compounds covered by the work of Werner, and helps to simplify the theory of such compounds."

It is of interest to note the following types,* which are the simplest of the series of silicic acids :—

Ortho-silicic acid $= SiO_2 \cdot 2H_2O =$

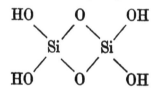

Meta-silicic acid $= 2SiO_2 \cdot 2H_2O =$

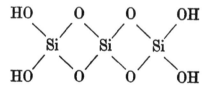

Tri-silicic acid $= 3SiO_2 \cdot 2H_2O =$

Tetra-silicic acid $= 4SiO_2 \cdot 2H_2O =$

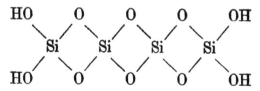

and so on to, perhaps, much longer-chain molecules.

* Taken from a paper on the " Origin of Primary Ore Deposits," by J. M. Campbell, *Bulletin of the Institute of Mining and Metallurgy* (October 1920); or see *Chem. News* (1920), 121, p. 278.

In applying the octet theory to the above series of compounds it is only necessary to draw half the atoms, as shown in Fig. 44,

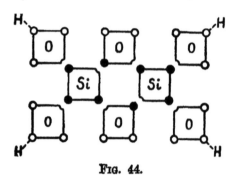

FIG. 44.

to fit them together so that the oxygen and silicon atoms share electrons : thus avoiding *in this case* a perspective drawing.

**Organic Compounds.**—Referring to the equation made use of in this theory, it will be found upon applying the formula to all organic compounds involving carbon, oxygen and hydrogen that they have valences in accordance with the ordinary theory. In each case a pair of electrons held in common corresponds to a bond. Two or three such pairs held between octets correspond to double and treble bonds respectively. In Table XVI a series of carbon compounds of the type $C_xH_{2y}$ is given together with the values $n$, $e$ and $p$.

TABLE XVI

| $C_xH_{2y}$ | $n$ | $e$ | $p$ | Constitution |
|---|---|---|---|---|
| $CH_2$ | 1 | 6 | 1 | Impossible |
| $CH_4$ | 1 | 8 | 0 | $CH_4$ |
| $CH_6$ | 1 | 10 | −1 | Impossible |
| $C_2H_2$ | 2 | 10 | 3 | $HC \equiv CH$ |
| $C_2H_4$ | 2 | 12 | 2 | $H_2C = CH_2$ |
| $C_2H_6$ | 2 | 14 | 1 | $H_3C - CH_3$ |
| $C_2H_8$ | 2 | 16 | 0 | $CH_4 + CH_4$ |
| $C_3H_2$ | 3 | 14 | 5 | Possible only in a ring |
| $C_3H_4$ | 3 | 16 | 4 | $H_2C = C = CH_2$ |
| $C_3H_6$ | 3 | 18 | 3 | $H_3C - \overset{H}{C} = CH_2$ ; or as a ring |
| $C_3H_8$ | 3 | 20 | 2 | $H_3C - CH_2 - CH_3$ |
| $C_3H_{10}$ | 3 | 22 | 1 | $HC_3 - CH_3 + CH_4$ |
| $C_4H_2$ | 4 | 18 | 7 | $HC \equiv C - C \equiv CH$ |
| $C_4H_4$ | 4 | 20 | 6 | $H_2C = C = C = CH_2$ ; or as a ring |

**Shapes of Molecules.**—Shapes of molecules can only be given diagrammatically by way of possible illustration, and much importance need not be attached to the following figures showing a few structures. The angularity involved in these representations obviously results from the arbitrary practice of drawing imaginary lines from electron to electron, or in delineation of the cubic symmetry involved. The following Figs. 45, 46 and 47 taken from Langmuir's paper will, however, be of interest :—

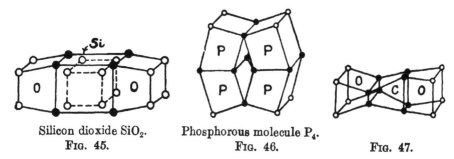

Silicon dioxide SiO₂.
FIG. 45.

Phosphorous molecule P₄.
FIG. 46.

FIG. 47.

In Fig. 45 the inner octet of the silicon atom is shown, but not the most central part. For a more detailed structure, see Figs. 24 and 25 (p. 95). It will be seen that these shapes involve the idea of different atomic volumes and the interaction of the atomic forces which give rise to ' distortions ' as illustrated. The carbon dioxide molecule is also of interest, as shown in Fig. 47.

**Atomic Volumes.**—It is well known that the atomic weight of an element multiplied by the volume occupied by a unit mass of the element is the *atomic volume*. The volume occupied by a unit mass is called the *specific volume* and it is inversely proportional to the density of the element : therefore,

$$\frac{Atomic\ Weight}{Density} = Atomic\ Volume.$$

The volume of the atom is not, however, always a fixed quantity like atomic weight, but varies in different compounds owing probably to the interaction of their fields. There are nevertheless certain regularities. See in this connection Chapter VII.

In discussing the volumes of the ' cubic ' atoms, Langmuir says : " For the molecular volume of magnesium oxide it appears that the atomic volume of the octet is less than 5·5. It is reasonable to suppose that with larger forces it might approach a limiting value of about 4·0. The volume of a single neon octet is $6·2 \times 10^{-24}$ c.c., corresponding to a cube having an edge of $1·9 \times 10^{-8}$ cm."

According to the views advanced by Langmuir, "the large atomic volumes of the alkali metals are due to the volumes occupied by single free electrons rather than to a large volume of the atom as such."

**Magnetic Properties of Certain Elements from the Point of View of the Octet Theory.**—The structures of iron, cobalt and

9

nickel differ from those of the elements preceding them, as in the atoms of these well-known magnetic elements there are 24 electrons arranged at the corners of 3 concentric cubes. Since these electrons are probably held in these positions by *magnetic* forces (see postulate 5), their arrangement in 4 concentric layers fits in with Parson's * theory of the cause of magnetism in iron; while A. W. Hull † in studying the structure of iron as revealed by X-ray analysis found that the atoms were arranged in a centred-cubic-lattice pattern— that is, each atom is surrounded by 8 others, these taking up positions on diagonals extending from the central cube. Hull also found that the intensities of the lines of the X-ray spectrum of iron were accounted for on the assumption that 8 of the 26 electrons in each atom were themselves arranged along the ' cube diagonals ' at a distance from the centre of the system equal to one-fourth the distance to the nearest atom; so that if all the electrons are in displaced positions along cube diagonals in 4 groups of 2, 8, 8 and 8 at distances $\frac{1}{32}$, $\frac{1}{16}$, $\frac{1}{8}$ and $\frac{1}{4}$, respectively, of the distances to the nearest atom, all the observed facts are well accounted for. Thus Hull's theory agrees with the octet theory as to the structure of the iron atom, except that the latter theory, as postulated, requires the radii of the electron shells to be more uniform than the values deduced by Hull. Nickel has a similar construction, according to Hull, but he has not determined the probable positions of the electrons.

Other conditions are given by Langmuir, which may be enumerated as follows :—

1. Ferromagnetic properties are also conditioned by the arrangement of the atoms with respect to one another.

2. In a pair of adjacent iron atoms, according to Hull, there are 6 electrons in line between the centres of the pair.

3. The outer electrons are probably held by weaker constraining forces, while the inner ones are held as firmly in position as those in·the atoms of argon.

4. Thermal agitation would probably disturb the outer electrons and thus destroy the magnetic properties if of sufficient intensity, though the heat may not be sufficient to destroy the crystalline form.

5. The Heusler alloys which consist of manganese and copper, together with small amounts of aluminium, arsenic, etc., owe their strong magnetic properties to the manganese which derives the extra electrons needed, to simulate iron and give rise to the magnetic state, from the copper present. Other elements than copper, such as phosphorus and nitrogen, serve the same purpose as copper and give paramagnetic alloys.

6. Similarly, chromium and vanadium form similar alloys, but they are magnetically weaker.

7. " In all these cases it is probable not only that the outer shell of the manganese, chromium, or vanadium atom takes up

* *Smithsonian Inst. Pub. Misc. Coll.* (1915), 65, No. 11.
† *Phys. Review* (1917), 9, p. 84.

electrons to revert to iron, but that atoms arrange themselves in a crystal structure which helps to bring out their magnetic properties (perhaps always the centred-cubic lattice)."

8. The conditions that might obtain in Hadfield's manganese steel which is practically non-magnetic are not mentioned by Langmuir.

9. The sharp break in the magnetic susceptibility between nickel (very high) and copper (very low : almost zero) is in accord with this theory. In metallic copper and in cuprous salts the electrostatic forces predominate, as revealed in the determination of the positions of the electrons, while in cupric salts the magnetic forces play an important part, these salts being rather strongly paramagnetic. The bivalent cupric ion corresponds to the bivalent ions of Fe, Co and Ni, all having paramagnetic properties; while the cuprous ion is diamagnetic and corresponds to $\beta$-nickel (see Table XI, p. 101).

10. The fact that argon has a full complement of electrons arranged at the corners of a cube does not lead to paramagnetic properties, as not only the general positions of the electrons have to be taken into account, but the revolution of the electrons, and in this case the resultant magnetic moments of the revolving electrons are zero, as argon is highly diamagnetic (see $H_2$-molecule above). Langmuir remarks : "It is only when the electrons are under much weaker constraints, but yet are held by magnetic rather than electrostatic forces that we should expect paramagnetism."

The magnetic properties of some of the elements of the rare-earth group are discussed in the section dealing with these elements.

**Isomorphism, Isosterism and Covalence.**—In a paper with this title, Langmuir * develops the octet theory further, but as the principle of isosterism has been given in the previous chapter—as applied to molecules (isosteric compounds) only—the application of the principle to groups of atoms which hold pairs of electrons in common and to ions will be given here. Isomorphous crystalline substances will be briefly reviewed in connection with the theory.

Just as there are what might be termed *fixed* molecules, which are isosteric, like $N_2O$ and $CO_2$, so too there are groups of atoms or single atoms which have electrically charged states (radicles or ions) which may be co-ordinated from the point of view of the octet theory. These substances are arranged in types, as will be seen from Table XVII on the following page.

It is to be noted that the same number and arrangement of electrons is the basis of isosterism ; and a ' comolecule ' is a group of atoms held together by pairs of electrons shared by adjacent atoms ; and these may, therefore, be isosteric if they contain the same number and arrangement of electrons.

It must be remembered, in examining the table, that the signs

* *Am. Chem. Soc. Journ.* (1919), 41, p. 1543.

after the symbols mean loss of electrons if they are $+$, so that $Na^+$, for example, would correspond to Ne, as would $F^-$: but the last-named atom has *gained* an electron. Similarly, $CH_4$ corresponds to $NH_4^+$. When co-molecules have the same charge they are said to be isoelectric; e.g. $N_2$ and CO are isoelectric. $N_3^-$ and $NCO^-$ are also isoelectric.

"The essential differences between isosteres are confined to the charges on the nuclei of the constituent atoms." Thus the nuclear

TABLE XVII

| Type | Isostere |
|------|----------|
| 1 | $H-$, He, $Li+$ |
| 2 | $O--$, $F-$, Ne, $Na+$, $Mg++$, $Al+++$ |
| 3 | $S--$, $Cl-$, Ar, $K+$, $Ca++$ · |
| 4 | $Cu+$, $Zn++$ |
| 5 | $Br-$, Kr, $Rb+$, $Sr++$ |
| 6 | $Ag+$, $Cd++$ |
| 7 | $I-$, Xe, $Cs+$, $Ba++$ |
| 8 | $N_2$, CO, $CN-$ |
| 9 | $CH_4$, $NH_4+$ |
| 10 | $CO_2$, $N_2O$, $N_3-$, $CNO-$ |
| 11 | $NO_3-$, $CO_3--$ |
| 12 | $NO_2-$, $O_3$ |
| 13 | HF, $OH-$ |
| 14 | $ClO_4-$, $SO_4--$, $PO_4--$ |
| 15 | $ClO_3-$, $SO_3--$, $PO_3--$ |
| 16 | $SO_3$, $PO_3-$ |
| 17 | $S_2O_6--$, $P_2O_6----$ |
| 18 | $S_2O_7--$, $P_2O_7----$ |
| 19 | $SiH_4$, $PH_4+$ |
| 20 | $MnO_4-$, $CrO_4--$ |
| 21 | $SeO_4--$, $AsO_4---$ |

charges in terms of electrons for $CO_2$ are 6 for C and 8 for O, so that together there are $2\times8+6=22$ electrons in the molecule. Similarly, in $N_2O$ the figures are $2\times7+8=22$; and these two compounds are remarkably alike, as already shown.

Considering now crystallographic data the isomorphism of cyanides and trinitrides affords an example, since the octet theory indicates similar structures, thus—

$$K^+(N=C=O)^- \text{ and } K^+(N=N=N)^-$$

The covalence of potassium is zero, and nitrogen and oxygen are bivalent, while in the trinitride one N is quadricovalent, the other being bicovalent.

Diazomethane should, according to this theory, have the structure

$$H_2C=N=N,$$

and this compound should be isosteric with

$$H_2C=C=O,$$

but the latter compound is not given in Beilstein; it should closely resemble diazomethane in all its physical properties, such as freezing-point, vapour pressure, viscosity, etc. Diazomethane is very poisonous, and the question might be raised as to whether this isostere if formed will have toxic properties.

Referring to the electric charges, Langmuir says : " No direct comparison can be made of the physical properties of isosteres having different electric charges. Thus we should not expect sodium salts to resemble neon, even though the sodium ion is an isostere of the neon atom—the electric force round the ion is sufficient to account for the differences in physical properties. . . . It is evident [however] that if any two substances are very much alike in physical properties, then any isoelectric isosteres of these substances should show similar close relationships with one another " : e.g. argon and nitrogen ($N_2$). See Table XVII : Types 3 and 8. " Therefore, the chlorine ion is isoelectric with argon and should have a close resemblance to the cyanogen ion which is isoelectric with nitrogen. The striking similarity of chlorides and cyanides is thus directly correlated with that between argon and nitrogen."

Summarising this paper, the octet theory indicates that compounds with the same number of atoms and electrons may have the same electronic arrangement, in which case they are isosteric, and resemble each other closely in physical properties. Table XVII gives molecules, ions and atoms, which are isosteric, these being grouped in 21 types.

In cases where the isosteric groups have the same charges they are said to be isoelectric and their properties are directly comparable, as, for example, $N_2$ and $CO$; but when the charges are unlike, the similarity may manifest itself between properly chosen compounds, as, for instance, sodium nitrate and calcium carbonate, and these substances have similar crystalline forms.

The following cases of crystalline isomorphism are predicted by this theory and they are found to exist as such, the parallelism being evident. Selected examples are given in Table XVIII.

TABLE XVIII

| Type | | |
|---|---|---|
| 2 | NaF . . . . . . . | MgO |
| 10 | $KN_3$ . . . . . . . | KNCO |
| 11 | $KNO_3$ . . . . . . | $SrCO_3$ |
| 14 | $KClO_4$ . . . . . | $SrSO_4$ |
| 14(b) | $NaHSO_4$ . . . . | $CaHPO_4$ |
| 21 | $MnSeO_4 \cdot 2H_2O$ . . . | $FeAsO_4 \cdot 2H_2O$ |
| | etc. etc. | |

The following list contains hypothetical compounds which should, if found to exist, show the relationships with known compounds :—

| | | |
|---|---|---|
| 2 | $MgF_2$ . . . . . . | $Na_2O$ |
| 3(b) | $K_2S$ . . . . . . | $CaCl_2$ |
| 15 | $NaClO_3$ . . . . . . | $CaSO_3$ |
| 15(b) | $KHSO_3$ . . . . . . | $SrHPO_3$ |
| | etc. | |

Argon is isosteric with the potassium ion; just as methane is isosteric with the ammonium ion; from which it follows that the potassium and ammonium ions have similar properties, especially since argon and methane resemble each other physically. Relations of this kind lead to the classification as shown in Table XVII. From these classifications more complex bodies are grouped, as instanced by cyanogen and chlorine ions leading to the similarity in the solubilities of nitrates and perchlorates; and again, carbonates and sulphates should be related, as appears to be the case in the examples given.

From the experimental data studied, the crystalline form of a compound depends upon the structure as revealed by the octet theory. In particular, the experimental data justify the following conclusions (taken from Langmuir's paper cited above) :—

1. The covalences of sodium, potassium, chlorine in chlorides, are zero.

2. The covalence of the central atom is 4 in nitrates, carbonates, sulphates, perchlorates, phosphates, permagnates, chromates, selenates, arsenates, bromofluorides, etc.

3. Carbonates and sulphates are not isomorphous, the covalence of the central atom being 4 and 3, respectively.

4. Nitrates and chlorates are not isomorphous, the covalence of the chlorine being 3 in chlorates.

5. The applicability of the octet theory to complex inorganic compounds receives further confirmation by its ability to explain such cases of isomorphism as between $Na_2BeF_4$ and $MnCl_2 \cdot 4H_2O$; $K_2SO_4$ and $ZnI_2 \cdot 4NH_3$; $K_2SnCl_4 \cdot 2H_2O$ and $K_2FeCl_5 \cdot H_2O$; $NaAlSi_3O_8$ and $CaAl_2Si_2O_8$, etc.

# CHAPTER XVI

## CO-ORDINATING THE RUTHERFORD AND LEWIS-LANGMUIR THEORIES

THE above title is perhaps not, strictly speaking, accurate ; for, so far as the writer is aware, the two theories indicated have not been officially co-ordinated. It is of interest, however, to discuss the possible union of ideas in this case.

If the discoveries and views of Rutherford (Chapter VIII) are placed parallel with the views given by Langmuir (Chapter XIV) it will be seen that there is close correspondence between them. Langmuir's nitrogen atom, however, does not quite agree with Rutherford's, if in the former's theory the kernel of nitrogen should be the *helium* atom ; whereas, according to the latter's theory based upon direct experiment, the central part should consist of two hydrogen units, as shown in one of the diagrams in the concluding part of Chapter VIII. There is, nevertheless, agreement as regards the positive charges of the kernel or central part.

However, from Rutherford's and Langmuir's ideas and the former's experiments it is possible to construct diagrams of the oxygen and nitrogen atoms which are shown by Figs. 48 and 49. The data to the right of each figure perhaps serve to indicate the possibilities of such structures better than any extended argument.

It should be realised that certain fundamental facts are in the process of co-ordination, and just what the final adjustments will be, to make the mechanism a complete working model cannot be foretold with certainty. This remark applies generally to the atomic theories recorded and discussed throughout these pages.

The accompanying Figs. (48 and 49) will help to make some of these ideas clearer.

Now to make the atoms *electrically cohesive structures*, if such an expression may be used, all the + and − quantities must be linked up, so that the principle, if true in the sharing of electrons in the Lewis-Langmuir molecules, may also be true of the complex atomic nuclei, as they may possibly share electrons amongst themselves, thus causing the whole system to be linked up cohesively : such at any rate are the ideas these two theories suggest to the writer ; see, however, Chapters VI and VII, since the idea of an internal ionic state is suggested in the case of the larger atoms. The rotation of the electrons, perhaps in pairs, may supply the necessary force to bring about the complete system of internuclear linkages. Magnetic as well as electrostatic forces are probably involved.

The divergence of the two theories as regards the nitrogen atom may be more apparent than real. If the electrons in the polar axis of the nitrogen atoms can under certain circumstances

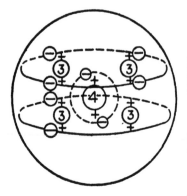

OXYGEN ATOM.

Mass $= 4 \times 3 + 4 = 16$.
2 resultant $+$ charges.
Valence $= 2$ (will share 2 electrons
            to complete the octet).
Atomic number $= 8$.
Electrons in circles $= 8$.

FIG. 48.

go to the outer 'corners,' the innermost nucleus or kernel is robbed of some of its stability, so that such a nucleus might break up on being struck with a high-speed missile, such as a fast-moving $\alpha$-particle, as ejected from radium C when disintegrating. Conse-

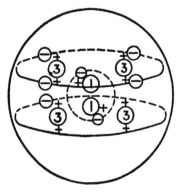

NITROGEN ATOM.

Mass $= 4 \times 3 + 2 = 14$.
3 resultant $+$ charges.
Valence $= 3$ (will share 3 electrons
            to complete the octet).
Atomic number $= 7$.
Electrons in circles $= 7$.

FIG. 49.

quently parts of the kernel may be driven out of the atom, as appears to be the case in Rutherford's recent experiments (see Chapter VIII). In these experiments masses of 3 were also dislodged and these in relatively greater numbers than those of mass 1;* and the former may be the components of the octet system lying between the 'corner-electrons,' as shown in Figs. 48 and 49.

Thus it will be seen that five features are brought into con-

* The actual numbers were very small.

sideration by taking into account those two theories conjointly, which may be summarised as follows :—

(i) That the central part of the atom may not always be a helium atom, but it may be what might be termed an isotopic equivalent. It is probable that the Lewis-Langmuir theory can be interpreted in this manner.

(ii) That the electrons in the kernel may under certain conditions go to the outer regions of the atom ; or that they may be mobile. This conception seems to be covered by the above theory, especially by the view put forward by Lewis.

(iii) That the principle of sharing electrons with positive nuclei may be extended to the nuclear parts within the atom. This is a somewhat obvious deduction by analogy.

(iv) That the electrons may revolve, possibly in pairs, and thereby give rise to directive forces (electromagnetic) necessary to link up the electrons and nuclei as a complete system more capable of resisting change in consequence. This view appears to be gaining ground and it has been considered by Langmuir,[*] Oxley [†] and others. In fact, the tendency seems towards a Rutherford-Bohr-Lewis-Langmuir atom.

(v) That the structure of some of the atoms may be somewhat as shown by Figs. 48 and 49.

NOTE.—The large parallel ' circles ' in Figs. 48 and 49 are drawn to show the positions of the electrons in space : to give perspective to the drawing.

It must be remembered that it is a very considerable climax of experimental achievement to have provided facts which enable deductions to be made, as contained throughout these pages.

[*] *Nature* (1920), 105, p. 261.
[†] *Ibid.* (1920), 105, p. 327 ; see **Fig. 39A above.**

# CHAPTER XVII

## THE LEWIS-LANGMUIR OCTET THEORY OF VALENCE APPLIED TO ORGANIC NITROGEN COMPOUNDS *

In the previous chapters reference is made to the possibility of the more central electrons going to the outer polar regions (or to corner positions) of the atom in certain compounds, particularly those of nitrogen. In order to be perfectly impartial in this exposition of the subject and not to impute to the authors of this theory ideas which they may have good reason to repudiate, the following brief description of certain expansions in the theory as applied to organic nitrogen compounds will be given.

There is no reference in this extension of the theory to the particular electronic movement, or displacement, as indicated in the previous chapter, but it is suggested by Langmuir that certain electrons may move into what might be termed edge-centred positions, very much as atoms in certain crystals are now known to occupy face-centred or central positions in the space-lattice. It will be desirable to review slightly the original theory.

Reasons have been given in the earlier papers that when the full-octet complement of 8 electrons is not shared by other atoms they arrange themselves at the eight corners of a cube. A pair of electrons held in common by two atoms, however, act as though they were located at a point.

Electrons in an atom having a covalence of 4, involving 8 electrons in its octet, are, according to this view, drawn together into 4 pairs which arrange themselves symmetrically at the corners of a regular tetrahedron and, as Lewis has already pointed out, by this means stereoisomerism of carbon compounds is fully explained.

The same reasoning applies to atoms of nitrogen, phosphorus, sulphur, etc., whenever these atoms are quadricovalent, so that stereoisomers are obtained when the structural conditions imitate those of carbon compounds. The experimental facts concerning these substances support this conclusion.

When only three pairs of electrons of an octet are shared an arrangement, as shown by Fig. 50, is suggested. The three pairs of electrons, A, B and C, which correspond to the covalence bonds are arranged in space approximately like the three corners

* See *Am. Chem. Soc. Journ.* (1920), 42, p. 274.

of a regular tetrahedron, the fourth corner corresponds to the unshared pair of electrons D.

This arrangement of the 3 bonds in space agrees exactly with Hantzsch and Werner's * theory, which has so well explained isomerism of—

Aldoximes,
Ketoximes,
Hydrazones,
Osazones,
Diazo-compounds, etc.

Moreover, P. Neogi's † recent theory involving such substances assumes an identical arrangement of the bonds, as indicated.

Continuing the subject, the octet valence theory is successfully applied to organic nitrogen compounds, as already partly indicated, and inorganic nitrogen compounds and salts are discussed in the more recent contribution with which this chapter deals.

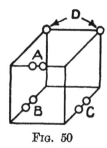

FIG. 50

The number of available electrons in the outside shell of any atom is supposed to be given by a number corresponding in many cases to the Group Number in the ordinary Periodic Table (see Tables XI and XII, pp. 101, 102).   Representing this number by B, it frequently corresponds to the maximum positive valence as in the ordinary valence theory, and is—

1 for sodium,
4 for carbon,
5 for nitrogen,
6 for oxygen, and
7 for chlorine.

Thereby it is shown that the octet theory is in agreement with the ordinary valence theory whenever the ordinary formulæ are based on unity for hydrogen and 8—B for each other element, so that C=4, N and P=3, O and S=2, Cl, Br and I=1.

On the other hand, in all formulæ in which the valences differ from these a modification is required to agree with the octet theory. The octet theory is applied to the following compounds, the formulæ of which require such modification.

* Berichte (1890), 23, p. 11 ; Chem. Soc. Abs. (1890), 58, p. 348 ; see also New Ideas in Organic Chemistry, by A. Werner, translated by E. P. Hedley (1911).
† Am. Chem. Soc. Journ. (1919), 41, p. 622.

Sodium chloride : $Na^+Cl^-$ (the covalence of both atoms is zero).

Ammonium chloride :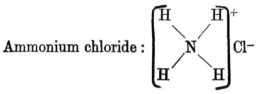

(the nitrogen in this compound is quadricovalent).

Triphenylmethyltetramethyl :

$$[N(CH_3)_4]^+ [(CPh_3]^-$$

In this compound the nitrogen is quadricovalent, while the central carbon atom in the anion ($-$) is tercovalent.

Diazophenol :

$$R{<}^{O}_{N=N}$$

Diazonium compounds :

$$[R-N\equiv N]^+ [OH]^-$$

or,

$$R-N=NOH$$

Triazo-compounds :

$$R-N=N=N$$

Hydroxylamine :

$$H_2NOH$$

or,

$$H_3N-O, \text{ etc.}$$

According to this theory all the salts are completely ionised before they enter into solution. This conception is in accord with the work of Milner, Ghosh and others : see Chapter XV. It explains why there are weak acids and weak bases, but no weak salts.

The known cases of isomerism, including those of stereoisomerism of nitrogen, phosphorus and sulphur compounds are in full accord with the octet theory.

The fact that organic cyanates, cyanides and nitrites exist in two isomeric forms, while the corresponding inorganic salts exist in only one form is explained, since the nulcovalent atoms of the metals in the inorganic compounds are not attached to definite atoms of the acid radicles. The available data on phosphonium, arsonium, sulphonium and oxonium compounds are in full accord with this theory, which gives for these compounds constitutions closely resembling those previously given by Werner. The writer has drawn freely from the original in presenting the foregoing views.

# CHAPTER XVIII

## THE OCTET THEORY AS A CLUE TO CERTAIN TYPES OF MOLECULAR INSTABILITY, AND A FEW REMARKS

THE octet theory has been successful in elucidating certain kinds of atomic and molecular stability, but the views here recorded are not necessarily final as regards detail, assuming the general principles to be sound. For example, the disposition of electrons in the free oxygen and nitrogen atoms may be as shown by Figs. 51 and 52, so that in combination, such as $O_2$ (see Chapters XIV and XVI), the oscillation of the kernel might tend to weaken a pair of corner-electron bondages in this case; whereas with $N_2$ the oscillation of

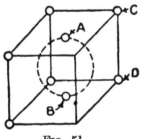

FIG. 51.

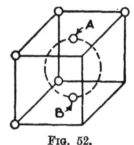

FIG. 52.

the kernel, or its electrons, could only weaken one corner bondage at a time, assuming that one central electron of each atom respectively goes to a corner; therefore the latter is a more stable molecule than the former; and this is a well-known fact. To make this clear, referring to Fig. 51, if A and B in oscillation move close up to C and D the latter pair would have their attraction momentarily weakened by the intrusion of a like pair, since if they both occupied equipotential places (side-by-side positions) the attraction per pair would be halved.

Extending these ideas, one can almost see why catalysts influence reactions without taking direct part in them. Suppose molecules, or molecular parts, come into close contact with a metallic surface or substance for example, so that at the moment of contact the sharing electrons at the C—D corners have their bondage or affinity weakened, and assume the temperature to be such as already to overstrain the bondage (owing to the vibration of the system), this weakening might just prove sufficient for the rupture, or change-over, to take place whereby other atoms combine with those so influenced, or a new polymeride of an existing combination is formed.

Passing to another most interesting chemical phenomenon, one can begin to understand the rationale of unstable compounds such as certain peroxides, these being formed by oxygen atoms attached to one another by sharing electrons, as shown by Figs. 53 and 54.

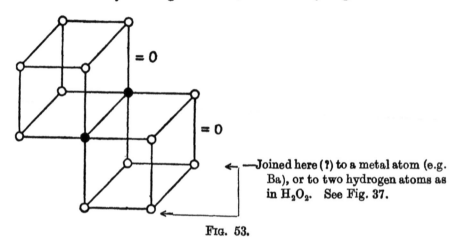

= O

= O

← —Joined here (?) to a metal atom (e.g. Ba), or to two hydrogen atoms as in $H_2O_2$. See Fig. 37.

FIG. 53.

This type of union (Fig. 53) *with hydrogen, barium, etc.*, is, however, contrary to Langmuir's theory, since the *hydrogen* atoms do not, strictly speaking, share electrons (see p. 109); but in conformity with the equation

$$p = \tfrac{1}{2}(8n - e)$$

there is only one pair shared and that is between the oxygen atoms, as shown. As already explained—

p = pairs formed (electrons shared).
n = octets formed or completed.
e = total number of available electrons in the shells of the atoms forming the given molecule.

Therefore, with $H_2O$ * $p = 0$, $H_2O_2$ $p = 1$.

In the case of hydrogen peroxide ($H_2O_2$) the model would be as depicted by Fig. 54 (or see Fig. 55 for an alternative arrangement of H nuclei, as one may not be justified in assigning such definite positions to the electrons). Each hydrogen nucleus (H), that is to say, the atom which has parted with its normally attendant electron, probably takes up a position between electrons somewhat as shown in the structures illustrated in Chapter XIV.

Passing to a subject of great potential interest, it will be remem-

* Not, of course, a peroxide.

bered that J. N. Collie and H. S. Patterson * apparently obtained
neon by passing an electric discharge through certain gases and
bombarding calcium fluoride with cathode rays. In one of these

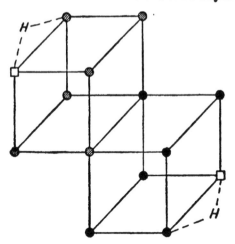

ELECTRON FROM HYDROGEN = □
LOWER OXYGEN ELECTRONS = ●
UPPER OXYGEN ELECTRONS = ◐

Fig. 54.

vacuum-tube experiments glass-wool was substituted for the
calcium fluoride, but neon still made its appearance. When the

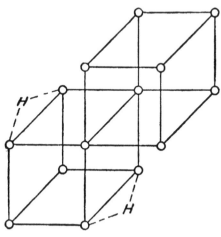

Fig. 55.

glass-wool was finely powdered, washed with chromic acid and
fused in an exhausted tube, no neon was given off. Other experiments † show the presence of neon under conditions which make it

* *Chem. Soc. Trans.* (1913), 103, p. 419.
† See A. W. Stewart, *Recent Advances in Physical and Inorganic Chemistry*
(1919), for a short account of the experiments apparently showing the *evolution*
of helium and neon.

almost appear that it was synthetically produced out of hydrogen by the action of the electric discharge. See in this connection, Chapter XX. Other investigators (see list below) have furnished conflicting evidence, so that the matter is not yet cleared up.

The following summarises the present position :—

1. Calcium fluoride was bombarded with cathode rays, when various gases were evolved, namely, hydrogen, oxygen, carbon monoxide and traces of silicon fluoride, together with a small quantity of neon. This shows that from the point of view of using any such results as a criterion, the substance was not pure.

2. As above, but with glass-wool in place of the calcium fluoride, a small quantity of neon was found.

3. Elaborate experiments were made to test whether the neon from the air had leaked into the tube, the conclusion being that leakage could not have taken place.

4. Experiments were made to prove that the presence of neon was not traceable to the solid substance bombarded in the tube, for the glass-wool when powdered and fused in a vacuum did not give off neon.

5. Aluminium electrodes were used in the experiments. When the aluminium was fused in a vacuum no neon was given off, *though hydrogen was evolved*.

6. R. J. Strutt (now Lord Rayleigh),[*] T. R. Merton,[†] A. C. G. Egerton,[‡] and A. Piutti and E. Cardoso [§] obtained results which did not confirm those above; but I. Masson's [||] and Sir J. J. Thomson's [¶] experiments in some respects supported them. See also further experiments by Collie.[**]

Troost and Ouvrard in 1895 claimed to have combined argon with magnesium ; the latter element was in a vapour state when the combination was thought to have been effected ; but Rayleigh and Ramsay and also Moissan in the same year failed to form any compounds by passing the electric discharge through argon with other elements present. Magnesium, of course, combines with nitrogen, and it is possible that in some of the early experiments nitrides were formed from traces of nitrogen present.

[*] *Roy. Soc. Proc.* (1914), 89, p. 499.
[†] *Ibid.* (1914), 90, p. 549.
[‡] *Ibid.* (1915), 91, p. 180.
[§] *Journ. Chim. Phys.* (1920), 18, p. 81.
[||] *Chem. Soc. Proc.* (1913), 29, p. 233.
[¶] *Nature* (1913), 90, p. 645 ; see also *Rays of Positive Electricity* (1913), p. 122.
[**] *Roy. Soc. Proc.* (1914), 90, p. 554 ; also same (1914), 91, p. 30. See also paper on the disappearance of xenon in discharge tubes, *Roy. Soc. Proc.* (1920), p. 349.

The octet theory affords ample reason for not expecting the inert gases to combine in the ordinary sense of the term, but since in the case of neon there is undoubtedly a feeble stray field round the atom it would not be surprising if unstable compounds like ozone could be formed with this element and oxygen which under certain conditions might be sufficiently stable to become a source of contamination in experiments of the above kind.   The instability in this case should be due to the weakness of the molecular structure (see peroxides above), but it might be quite stable when condensed on surfaces or occluded by metals, etc.   Such a compound might, therefore, exist as a widely distributed impurity which the discharge would break up into its elements.   It would have a molecular weight equal to one of the argon isotopes (see Table II, p. 14).

A. W. Stewart * raises some interesting questions : " Why, for example, since we get $PCl_3$ and $PH_3$, should we not expect to produce $PH_5$ just as easily as $PCl_5$ ?   Why is $PH_4Cl$ stable when $PHCl_4$ has not been isolated ?

According to the octet theory chlorine could acquire the five electrons of phosphorus and thus complete all the chlorine octets. Similarly, hydrogen atoms could give up their three electrons to phosphorus and thus complete the octet of phosphorus.   While $PCl_3$ has only three completed chlorine octets, the phosphorus octet shares a pair of electrons with each chlorine atom, as will be seen when constructing a diagram.   Langmuir's formula gives the following values :—

$$n=4, \ e=26 \text{ and } p=3.$$

In the case of $PH_4Cl$ there should be ions as explained in Chapter XV : valence, co-ordination number and covalence.   (See ammonium chloride, p. 140.)

The importance of *symmetry* and *divisibility* in all kinds of fundamental atomic phenomena cannot be overestimated.   One might almost say *discontinuity and symmetry reign supreme over the physico-chemical universe.*

* *Recent Advances in Physical and Inorganic Chemistry* (1919).

# CHAPTER XIX

## A NOTE OF CAUTION

In considering the experimental data and the theories presented in these pages, the reader should be aware of the fact that speculative ideas play a part in the literature of the class here represented, and while some ideas take root and become established, others seem too improbable to claim even temporary recognition.

It frequently happens, however, that the most transcendental hypotheses are suggestively close to the truth. The following examples from the studies of the writer will illustrate some of the efforts which are partly supported by recent experiments, but which on the whole fail to come up to expectations.

In 1914 the writer * made an attempt to establish the following generalisation : That the chemical elements can be broadly classified thus—

> " I. Whole-number atomic-weight elements which have no associates differing from them in atomic weight.
> " II. Whole-number atomic-weight elements composed of, or containing, chemically non-separable associates of different atomic weight."

The term " associates " referred to isotopes, as in II.

Prof. Soddy † had remarked : " The question naturally arises whether some of the common elements may not, in reality, be mixtures of chemically non-separable elements in constant proportions, differing step-wise by whole units in atomic weight. This would certainly 'account for the lack of regular relationships between the numerical values of the atomic weights." The radio-atoms afford abundant evidence of the atomic weights differing by helium units of mass 4, according to the Rutherford-Soddy law of mass change (see Appendix VI).

In the writer's attempt referred to, chlorine was assumed to have isotopic atoms of masses 35 and 40 mixed together in such proportions as to give a mean value 35·454 ; that is to say, one of mass 40 to ten of mass 35. Aston's recent work seems to prove the general supposition (I and II), but the values for chlorine are 35 and 37 in a ratio of about 3 to 1 (see Chapter II).

* *Chem. News* (1914), 109, p. 169.
† *Chem. Soc. Proc.* (1911), 99, p. 72.

A further effort was made by the writer * in 1915 to account for the seemingly exact whole-number atomic weights of C, N, O and F, whilst elements on each side of these had fractions for the most part ; therefore, it was thought that they might be made up of associate atoms † or isotopes. The essence of this effort was represented by a cyclic scheme based on an idea advanced many years ago by Sir William Crookes,‡ which in the writer's treatment took the form of a series of loops as obtained in analysing magnetic hysteresis phenomena.§ The following diagram, Fig. 56, was given, in which the ordinates represent approximate atomic weights.

Referring to Table II of Chapter II, it will be seen that the non-isotopic characters of C, N, O and F are proved, and a number of isotopic elements on each side of these has been discovered by the improved positive-ray method of Aston.‖

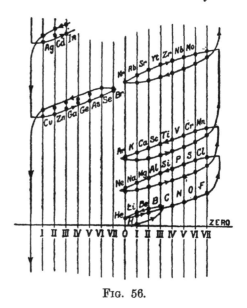

A still further attempt was made by the writer ¶ to formulate matters more fully : to evaluate the proportionate numbers of isotopes and to devise a tolerably regular scheme, which is shown by Table XIX (corrected in respect of nickel since first published).

The values 1, 2, 3 and 4, under letters *a, b, c* and *d,* were regarded at the time as ' outriders,' but they seem more allied with the central part of the atom in Ruther-

FIG. 56.

ford's and Langmuir's theories (see Chapters VIII and XIV). In the first column of figures *n* stands for a number of a series 1, 2, 3, 4, etc., and He represents the helium atom as a sub-atom or nuclear unit in the atomic edifice.

According to Rutherford's recent experiments, it looks as if the order should be reversed, or rather that 4 should be the multiplying number and ' *n* ' should be the mass unit to be multiplied. Whether this change should obtain throughout the table is of

---

* *Chem. News* (1915), 111, p. 157.
† Not to be confused with molecular association.
‡ See Presidential Address to the Chemical Society, 28 March, 1888.
§ See Ewing, *Magnetic Induction in Iron and other Metals,* last edition, p. 95.
‖ Lithium probably has two whole-number isotopes : see concluding note in Chapter VIII. In *Nature,* 23 June, 1921, Aston announces that nickel has isotopes of masses 58 and 60. Table XIX has been corrected in respect of one of the nickel isotopes.
¶ *Chem. News* (1920), 121, p. 105.

## TABLE XIX

| Element | $n \times$ He $a$ | $b$ | $c$ | $d$ | Isotopes and Prop. No. | | 'Calc.' and Exp. Mass. | |
|---|---|---|---|---|---|---|---|---|
| Li $=$ | $1 \times 4 +$ <br> $1 \times 4 + 1$ | | 3 | | $= 7$ <br> $= 5$ | | $3^3 \times 7$ <br> $1^3 \times 5$ $= 6 \cdot 93$ | $6 \cdot 94$ |
| Be $=$ | $1 \times 4 +$ <br> $1 \times 4 +$ | $2 +$ | | $4$ <br> $4$ | $= 8$ <br> $= 10$ | $(2^3 + 4^3)$ | $4^3 \times 8$ <br> $\times 10$ $= 9 \cdot 06$ | $9 \cdot 1$ |
| B $=$ | $2 \times 4 +$ <br> $2 \times 4 + 1$ | | 3 | | $= 11$ <br> $= 9$ | | $3^3 \times 11$ <br> $1^3 \times 9$ $= 10 \cdot 93$ | $10 \cdot 9$ |
| C $=$ <br> N $=$ | $2 \times 4 +$ <br> $3 \times 4 +$ | $\cdot\cdot$ <br> $2$ | | $4$ | | No Isotopes | $= 12 \cdot 00$ <br> $= 14 \cdot 00$ | $12 \cdot 00$ <br> $14 \cdot 008$ |
| O $=$ <br> F $=$ | $3 \times 4 + \uparrow$ <br> $4 \times 4 + 1$ | | $3$ <br> $\downarrow$ | | | | $= 16 \cdot 00$ <br> $= 19 \cdot 00$ | $16 \cdot 000$ <br> $19 \cdot 00$ |
| Ne $=$ | $4 \times 4 +$ <br> $5 \times 4 +$ | $2$ | | $4$ | $= 20$ <br> $= 22$ | | $4^3 \times 20$ <br> $2^3 \times 22$ $= 20 \cdot 22$ | $20 \cdot 2$ |
| Na $=$ | $5 \times 4 +$ <br> $5 \times 4 + 1$ | | 3 | | $= 23$ <br> $= 21$ | | $3^3 \times 23$ <br> $1^3 \times 21$ $= 22 \cdot 93$ | $23 \cdot 00$ |
| Mg $=$ | $5 \times 4 +$ <br> $6 \times 4 +$ | $2$ | | $4$ | $= 24$ <br> $= 26$ | | $4^3 \times 24$ <br> $2^3 \times 26$ $= 24 \cdot 22$ | $24 \cdot 32$ |
| Al $=$ | $6 \times 4 +$ <br> $7 \times 4 + 1$ | | 3 | | $= 27$ <br> $= 29$ | | $3^3 \times 27$ <br> $1^3 \times 29$ $= 27 \cdot 07$ | $27 \cdot 1$ |
| Si $=$ | $6 \times 4 +$ <br> $7 \times 4 +$ | $2$ | | $4$ | $= 28$ <br> $= 30$ | | $4^3 \times 28$ <br> $2^3 \times 30$ $= 28 \cdot 22$ | $28 \cdot 3$ |
| P $=$ | $7 \times 4 +$ <br> $8 \times 4 + 1$ | | 3 | | $= 31$ <br> $= 33$ | | $3^3 \times 31$ <br> $1^3 \times 33$ $= 31 \cdot 07$ | $31 \cdot 04$ |
| S $=$ | $7 \times 4 +$ <br> $8 \times 4 +$ | $\cdot\cdot$ <br> $2$ | | $4$ | $= 32$ <br> $= 34$ | | $4^3 \times 32$ <br> $\ddot{2}^3 \times 34$ $= 32 \cdot 06$ | $32 \cdot 07$ |
| Cl $=$ | $8 \times 4 +$ <br> $9 \times 4 + 1$ | | $\overset{\cdots}{3}$ | | $= 35$ <br> $= 37$ | | $\overset{\cdots}{3^3} \times 35$ <br> $1^3 \times 37$ $= 35 \cdot 50$ | $35 \cdot 46$ |
| A $=$ | $8 \times 4 +$ <br> $8 \times 4 +$ | | | $\overset{\cdots\cdot}{4}$ <br> $4 + 4$ | $= 36$ <br> $= 40$ | $(4^3 + 4^3)$ | $\overset{\cdots\cdot}{4^3} \times 36$ <br> $\times 40$ $= 39 \cdot 88$ | $39 \cdot 9$ |
| K $=$ | $9 \times 4 +$ <br> $10 \times 4 + 1$ | | 3 | | $= 39$ <br> $= 41$ | | $3^3 \times 39$ <br> $1^3 \times 41$ $= 39 \cdot 07$ | $39 \cdot 10$ |
| Ca $=$ | $9 \times 4 +$ <br> $10 \times 4 +$ | $\cdot\cdot$ <br> $2$ | | $4$ | $= 40$ <br> $= 42$ | | $4^3 \times 40$ <br> $\ddot{2}^3 \times 42$ $= 40 \cdot 06$ | $40 \cdot 09$ |
| Sc $=$ | $10 \times 4 +$ <br> $11 \times 4 + 1$ | | $\overset{\cdots}{3}$ | | $= 43$ <br> $= 45$ | | $\overset{\cdots}{3^3} \times 43$ <br> $1^3 \times 45$ $= 43 \cdot 50$ | $44 \cdot 1$ |
| Ti $=$ | $11 \times 4 +$ <br> $12 \times 4 +$ | $2$ | | $4$ | $= 48$ <br> $= 50$ | | $4^3 \times 48$ <br> $2^3 \times 50$ $= 48 \cdot 22$ | $48 \cdot 1$ |
| V $=$ | $12 \times 4 +$ <br> $12 \times 4 + 1$ | | 3 | | $= 51$ <br> $= 49$ | | $3^3 \times 51$ <br> $1^3 \times 49$ $= 50 \cdot 93$ | $51 \cdot 06$ |

TABLE XIX—(continued)

| Element | $n\times$He $a$ | $b$ | $c$ | $d$ | Isotopes and Prop. No. | 'Calc.' and Exp. Mass |
|---|---|---|---|---|---|---|
| Cr | $=\begin{matrix}12\times4+\\13\times4+\end{matrix}$ | $\overset{..}{2}$ | | 4 | $\begin{matrix}=52\\=54\end{matrix}$ | $\begin{matrix}4^3\times52\\ \ddot{2}^3\times54\end{matrix}\ =\ 52{\cdot}06 \quad 52{\cdot}0$ |
| Mn | $=\begin{matrix}13\times4+\\13\times4+1\end{matrix}$ | | 3 | | $\begin{matrix}=55\\=53\end{matrix}$ | $\begin{matrix}3^3\times55\\1^3\times53\end{matrix}\ =\ 54{\cdot}93 \quad 54{\cdot}93$ |
| Fe | $=\begin{matrix}13\times4+\\13\times4+\end{matrix}$ | 2 | | 4 | $\begin{matrix}=56\\=54\end{matrix}$ | $\begin{matrix}4^3\times56\\2^3\times54\end{matrix}\ =\ 55{\cdot}78 \quad 55{\cdot}84$ |
| Co | $=\begin{matrix}14\times4+\\14\times4+1\end{matrix}$ | | 3 | | $\begin{matrix}=59\\=57\end{matrix}$ | $\begin{matrix}3^3\times59\\1^3\times57\end{matrix}\ =\ 58{\cdot}93 \quad 58{\cdot}97$ |
| Ni | $=\begin{matrix}14\times4+\\14\times4+\end{matrix}$ | 2 | | $\overset{....}{4}$ | $\begin{matrix}=60\\=58\end{matrix}$ | $\begin{matrix}\overset{....}{4^3}\times60\\2^3\times58\end{matrix}\ =\ 58{\cdot}67 \quad 58{\cdot}68$ |

EXAMPLES

$3^3 \times 7 = 7 \times 27 = 189$
$1^3 \times 5 = 5 \times\ \underline{1 =\ \ 5}$
$28)\ \overline{194}$
$\underline{6{\cdot}93}$

$\ddot{2} = 1 + 1$
$\ddot{2}^3 = 1^3 + 1^3 = 2$
$\dddot{3} = 1 + 1 + 1$
$\dddot{3}^3 = 1^3 + 1^3 + 1^3 = 3$

NOTE.—The dots on the marginal line indicate that positive-ray experiments gave values the same as here shown. The cross indicates a one-unit difference between the values—see Table II.

course questionable—supposing the table to be near to the truth. All that can be said is that this suggestion is one that seems to fit in with the Lewis-Langmuir theory in a general way ; for, since each atom should have an electronic octet in its outer shell when perfectly stable, and since an octet may imply a quartet of double positive charges (see Chapters VIII and XVI) whether electrons are fully present or not, the multiplier, 4, is accounted for. Some of the figures are very large, but the limit appears to be 14, as the scheme in its present form seems to fail after nickel. Therefore, the large numbers may on this view stand for *radial-quartet members*, so that each radial spoke of the atom, so to speak, is made up to the masses represented in the first column of figures. Each ' spoke ' may be made up of separate units falling into place in each shell or shell-layer.

It will be seen that the calculated values are obtained by trial and error until values in the first column are found to give results close to the experimental ones. The dots over certain numbers indicate that the proportion is evaluated directly from their masses taken as a whole, and not from the cube of their masses, as in the other cases. The examples will explain the scheme. The dots may imply individuality as well.

It will be seen that the above table and Fig. 56 show a certain amount of regularity and agreement with experiment ; and that there is no violation of the oft-quoted coincidence of even and odd atomic weights with even and odd valences, except in the particular case of nitrogen, which is exceptional in this and in other respects. Notwithstanding the agreements with experiment, such schemes cannot be regarded as having more than a small element of truth in them. *They are not, as a matter of fact, founded upon a sufficient number of proved values to eliminate the element of uncertainty.*

*The examples of this chapter therefore serve to caution the reader not to accept as final fairly promising speculative schemes.*

As an example of the changes to be expected in schemes of the above kind, the early determination of the chemical atomic weight of boron was recorded as 11·0 ; a later determination revealed the value 10·90, but a still later one revealed the value 10·82. The isotopes of boron are known to be masses 10 and 11 ; and their mean value was stated by their discoverer to be 10·75 ± 0·07. Now in Table XIX it will be seen that if 2 is substituted for 1 to make up the isotope of mass 10, the mean value comes out at 10·77, which is very close to both the isotopic mean and the latest one determined chemically, the calculation being—

$$
\begin{array}{r}
11 \times 27 = 297 \\
10 \times \phantom{0}8 = \phantom{0}80 \\
\hline
35)\phantom{0}\phantom{0}377 \\
\hline
10\cdot77
\end{array}
$$

Thus it will be noted that the above Table is not wholly wrong in respect of boron, but an amendment becomes necessary.

It is desirable again to refer to Prout's hypothesis, since it is now obvious that all atoms are probably built up of mass units presumably of a value equal to or close to that of hydrogen. Hydrogen is not an exact whole number, since it has an appreciable fraction, seeing that the atom as a whole is so small in value. This would be better appreciated if the value were multiplied by 100, in which case it would have a very large decimal fraction, 0·8. This difficulty, as already explained, has been overcome by supposing that the mass is entirely electromagnetic, and when hydrogen condenses in a fundamental manner to form atoms of successive elements, the extra mass over that of the whole-number products is reduced by the interaction of electric fields. This is, of course, a highly speculative idea, like some of the foregoing ones of this chapter, and to many it will be difficult to accept such a view. The subject therefore bristles with difficulties, although great strides have been made in clearing away many obstacles, as will be seen from the experiments cited in these pages. The next chapter (XX) is of interest in connection with Prout's hypothesis.

# CHAPTER XX

## THE POSSIBLE COMPLEXITY OF HYDROGEN: ITS POSITION IN THE PERIODIC TABLE

ALTHOUGH Chapter X has been devoted to the structure of the hydrogen atom as deduced from an interpretation of its spectrum, it cannot be assumed that everything is known about this unique element. Hydrogen being the first member in the series (see Appendix I), and being so near to unity in atomic weight, special interest attaches to it. Its position in the periodic classification has been the subject of much discussion.

Aston has recently verified the mass of hydrogen by means of his improved positive-ray method (see Chapter II), its mass being in agreement with that obtained by chemical and density methods of determination. Aston remarks on the fact that it has not a whole-number atomic weight relative to all the other whole-number atoms thus far examined by this new method of analysis.

There are plausible arguments for placing this element either over lithium or over fluorine in the Periodic Table. With regard to the latter position, O. Masson * drew up a set of reasons favouring its position over fluorine. These are summarised in Newth's *Inorganic Chemistry*, substantially as follows :—

(i) It resembles fluorine in respect of its gaseous property.

(ii) It resembles the halogens in not being metallic.

(iii) It is unlike the alkali metals as it is diatomic ($H_2$), whereas these metals are monatomic (see Appendix VII).

(iv) It is readily substituted by Cl, Br or I in organic compounds.

(v) It would lead to six blank places in the Periodic Table if placed in Group I, involving the existence of elements having atomic weights between 1 and 4 (4=helium).

(vi) It would involve a difference between $H=1$ and $Li=7$ of 6 units, whereas if placed over F the difference would correspond better with the difference between F and Cl, which is 16.

---

* *Chem. News* (1896), 73, p. 283.

From this line of reasoning Masson concluded that hydrogen should be placed in the halogen group (VII) over fluorine.

There is, however, another argument that leads to the position over lithium, this being equally satisfactory. Referring to a recent work, *Text-Book of Inorganic Chemistry* (1917), vol. i, by J. N. Friend, H. V. F. Little, W. E. S. Turner and V. A. Briscoe, p. 274, the following statement appears : " It is generally agreed that the valency of hydrogen is unity, and this opens up two possible positions for the element in the table, namely, at the head of either the first or the seventh group. In accordance with its low boiling-point, and the diatomic nature of its gaseous molecule, many chemists prefer to include hydrogen in the seventh group along with the halogens ; and this is apparently justified by the fact that solid hydrogen bears no resemblance whatever in its physical properties to the alkali metals—on the contrary, it is typically non-metallic. Furthermore, Moissan showed that in the metallic hydrides hydrogen behaves like a non-metal, inasmuch as the hydrides do not conduct the electric current, and hence it cannot be regarded as comparable with alloys. But when the chemical behaviour of hydrogen is considered, facts speak strongly in favour of its metallic nature. The most stable compounds of hydrogen, as of the metals generally, are those formed by union with non-metallic elements. Thus, for example, although the halogens exhibit but little tendency to combine amongst themselves, save in the case of the two extreme elements, fluorine and iodine, yet they yield very stable compounds with hydrogen and the metals. The same truth applies with more or less completeness to the other non-metallic elements known. Consequently it seems most natural to regard hydrogen as analogous to a metal, and as such it is a more fitting forerunner of the alkali metals than of the halogens."

The only fair inference to be drawn from these two views is that hydrogen is eligible for *both* places in the table, and some chemists allocate this element accordingly.

There are thus *two* peculiarities about hydrogen :

1. It is not a ' whole-number ' element.
2. It is eligible for *two* places in the Periodic Table.

Now there is a *third* peculiarity arising from the present-day view, that all the elements are composed of some simple unit-element of a primary character. Helium (atomic weight=4) is known to be a product of radio-active change, and Rutherford has detached from N, O and C elementary masses which appear to have the mass values 1, 2 and 3 (see Chapter VIII), and this bears out the view that all the elements are built up of such units as represented by the masses 1, 2, 3 and 4. Prout's hypothesis that all the elements are fundamental polymerides of hydrogen, hydrogen being taken as unity, is now gaining acceptance, and it will be seen that the whole-number atomic weights of the elements and the work of Rutherford lend strong support to this view advanced over one hundred years ago.  Constructing all the elements out of a unit of the value 1·008 would, however, lead to many fractional values, and this would not be in harmony with Aston's findings (see Chapter II).  To meet this difficulty the idea of a contraction of mass, as the units are added or condensed together in a fundamental sense, has been suggested, and this idea would possibly be in harmony with electromagnetic ideas if all mass *is* electromagnetic ; in short, electromagnetic contraction of mass has been postulated by Rutherford and others.*

In pursuance of this contraction idea, the writer † has suggested that there may be a systematic contraction, and a sort of electromagnetic multiplication table was evolved, which may be stated thus—

$$\text{If } 1 \times 1·008 = 1·008 \text{ and}$$
$$4 \times 1·008 = 4·000,$$
$$\text{then } 2 \times 1·008 = 2·004 \text{ and}$$
$$3 \times 1·008 = 3·002.$$

A difficulty, however, arises here ; for if the contraction is limited to the production of helium units, these units themselves do not appear to contract, since oxygen would be $4 \times 4·0000 = 16$ exactly ; but Rutherford has broken up the oxygen atom, but not the helium atom, which is consistent as far as it goes, since the masses driven out of the oxygen atom have a value of 3·1 or 3·0 (see Chapter VIII).  Moreover, if the contraction is so limited, why should so many elements be so very stable and resist disintegration by ordinary means, such as that of very high temperature ?  It should be borne in mind that it is assumed that the greater the contraction the greater the stability of the resulting element.  This idea of stability was advanced by Sommerfeld,‡ who referred to W. Lenz, who it appears had previously given expression to a similar conception.  There is evidently a difficulty here that requires clearing up ; and as this is connected with hydrogen as a unit it may be stated as a *third* peculiarity, namely—

3. It (hydrogen) is apparently capable of contraction in the formation of certain units, say, those of values 4·0000, 2·004 and 3·002, but the stability of the atom in general does not appear to be governed by these contractions, as should be the case.§

* Those who may wish to trace the origin of this idea should consult Comstock's paper published in the *Am. Chem. Soc. Journ.* (1908), 30, p. 683 ; or the *Chem. News* (1908), 98, p. 178.  See also Rydberg, *Zeit. Anorg. Chem.* (1897), 14, p. 66.  Other references are : Risteen, *Molecules and Molecular Theory* (1895) ; Rutherford, *Phil. Mag.* (1911), 21, p. 669 ; (1914), 27, p. 488 ; Kaufmann, *Göttingen Nach.* (8 Nov., 1901).
† *Chem. News* (1920), 120, p. 181.
‡ *Atombau und Spektrallinien* (1919), p. 538.
§ Unless, of course, these units are structural ones and contract in the process of atom building.

There is a *fourth* peculiarity, namely—

> 4. It (ordinary atomic hydrogen) is not known to polymerise to form any elements. It is true that certain experiments (e.g. those of Collie and Patterson given in Chapter XVIII) point in this direction, but the evidence does not appear to be conclusive.

Turning now to Table XII, on p. 102, it will be seen that there appears to be what might be termed a shortage of elementary material in the case of elements at the lower end of Groups I and VII (7 and 17 Langmuir's notation). This may be referred to as the *fifth* peculiarity, though as yet not associated with hydrogen.

Now by supposing that hydrogen itself is complex and consists of three parts, namely—

> 1. A whole-number part of mass . . . 1·0000 (*a*)
> 2. A fractional part of mass . . . . 0·0077 (*b*)
> 3. An electron of electromagnetic mass * . 0·00055 (*c*)

it is possible to remove at once all the five difficulties, as will be seen from the following statement of the main points discussed above (one or two additional deductions are introduced in this summary) :—

### SUMMARY

Hydrogen appears to be unique in not being a whole-number atom. The fractional part of its mass is relatively great. Hydrogen does not find a proper place at the heads of Groups I and VII. Its chemical activity is evidently such as not to enable it to form stable polymerides answering to known elements. There is a gap at the end of Group I that does not accommodate any radio-active element. Similarly, there are gaps below manganese which seem difficult to fill, and one of these would have to be filled by a radio-active element which seems improbable. On the hypothesis that all elements are polymerides of hydrogen, a difficulty arises owing to the fractional value, suggesting that the fundamental unit has a mass of 1 exactly. Considering hydrogen as a complex atom composed of mass parts $a=1·0000$, $b=0·0077$, and $c$ (an electron) $=0·00055$, $a$ and $b$ could possibly be more properly placed at the head of Groups I and VII respectively, and thus account for the very strong affinity between these parts, and thereby explain the stability of the hydrogen complex (*ab* or *abc*), whilst the respective parts alone might have very great affinity, especially $a$, which

---

* This mass is variable when considered in the electric sense, and when moving with a velocity just short of that of light its mass would be equal to that of the entire hydrogen atom. All bodies are supposed to have electromagnetic mass and consequently this mass is not constant when it moves at a very high speed.

could polymerise and form the whole-number elements as revealed by Aston's recent experiments. Moreover, constructing a modified Langmuir Periodic Table, the absence of elements in Groups I and VII would suggest a shortage of elementary material, so to speak ; but this shortage would be made up at the heads of these groups by introducing the entities *a* and *b* at these places. This procedure reveals a numerical regularity, as will be seen by the additions shown in Table XII. The complexity of hydrogen thus indicated would account for the fact that it is not fully eligible for a position at the head of either of the above-mentioned groups. The question of atomic numbers to be assigned to these hypothetical parts is of interest but cannot be discussed here. The fact that ordinary hydrogen is not known to polymerise to form the elements would be expected if it carries an auxiliary part which deprives it of this property.

## NOTE

It is important to distinguish between a process which, when it operates, follows a given law, and the evidence that such a process always produces something.

The law may be rigorously true, as in the case of the Moseley-van den Broek *atomic number sequence*; but, and this is the point to be noted, the *material* which was necessary for the production of the elements may have been at times insufficient (in some unknown way) to give expression always to the law; so that lacunæ may exist which represent very few elements formed, or, in rare instances, none at all.

# CHAPTER XXI

THE ENERGY OF THE ATOM: ELEMENTS IN THE SUN AND
THEIR IONISATION: ATOMIC ENERGY AND SOLAR
RADIATION: SUN-SPOTS, AURORA AND MAGNETIC
STORMS: HIGH-PRESSURE EXPERIMENTS OF SIR C. A.
PARSONS: HIGH TEMPERATURES OBTAINED BY EX-
PLODING WIRES ELECTRICALLY

PRIOR to the discovery of radio-active phenomena it was not
expected that the atom itself was an internal source of great energy
relative to its size; at least, this idea had not been discussed
scientifically, since there was no foundation fact on which to
establish such a theory.

It is now known that in the process of atomic disintegration the
energy liberated from the atom is enormously greater than in
chemical reactions when properly comparing equal amounts of
substance. Dr. Crowther * has worked out some values for radio-
active atoms which are made use of here.

When hydrogen and oxygen are exploded to form water the
heat liberated is $69 \times 10^3$ calories for 18 grammes, which is about
3830 calories per gramme of water formed.

Radium in equilibrium with its disintegration products gives
out $1\cdot4 \times 10^6$ ergs per second, or 118 calories per hour per gramme,
but its period of half-change is long.

When a gramme of radium emanation has completely disintegrated, the total energy liberated during the change would be
about $2\cdot5 \times 10^9$ calories when including the energies of the short-
lived products immediately following it, which have half-periods
of 3, $26\cdot7$, $19\cdot5$ and $1\cdot4$ minutes respectively. It is to be noted that
there is still an atomic mass of 210 left, as the energy given refers
only to that evolved during the breaking down of the atom from a
mass of 222 to one of 210 by the emission of 3 helium atoms.

The ratio of the energy of the short-lived emanation and its
products to that of water formation out of its elements becomes

$$\frac{2\cdot5 \times 10^9}{3\cdot8 \times 10^3} = 658,000.$$

The energy given out in this radio-active change, that is to

* *Ions, Electrons and Ionising Radiations* (1919), p. 249.

say, evolved from within the atoms, is thus about 600,000 times greater than the intense chemical action of water formation.

The formation of $Li_2O$ represents slightly more energy than in the case of water—when disregarding that absorbed in breaking up the molecules of hydrogen and oxygen initially—viz. $4.7 \times 10^3$ calories ; whilst lithium fluoride (LiF) has a heat of formation of about 110,000 calories per gramme molecule. This would correspond to 4230 calories per gramme. Thus it will be seen that any one of these most energetic reactions is very feeble compared with the energy of the atom.

It is of interest now to turn to the octet theory, as there is a certain, though somewhat remote, parallelism in the actions involved in what might be termed *internal* and *external* atomic phenomena, for Langmuir says : " The fluorine atom has 7 electrons in its shell. Its properties are therefore largely determined by the tendency to take up an additional electron to complete the octet. Thus, if the molecule is $F_2$, we have $n=2$; $e=14$, whence $p=1$. By sharing a single pair as shown in Fig. [37] both octets are completed. The very low boiling-point of fluorine indicates that the stray field round this molecule is small. On the other hand, the remarkable chemical activity shows that there is a strong tendency for these atoms to avoid sharing their electrons with each other. Thus, when lithium and fluorine are brought together, the extra electrons from the lithium atoms are taken up by the fluorine atoms, and thus each atom is able to get its own octet instead of being compelled to share it with another atom. The very large heat of formation of lithium fluoride (about 110,000 calories per gramme molecule) must be, in greater part, the heat equivalent to the difference between a free octet and one which shares two of its electrons. The energy liberated when an *atom* of fluorine, with its 7 electrons in the shell, takes up another electron would be greater than the above by the energy required to disintegrate fluorine into atoms—a quantity which has never been determined, but which must be very large.

" The question arises, why do atoms share their electrons with each other, if there is so much tendency for the octets to stay separate ? The answer, of course, is that there are not enough electrons in the outside shells of atoms to form octets round their kernels unless they share them with each other. Furthermore, most atoms if they completed their octets without sharing electrons would have very high charges on their atoms which would tend to prevent their formation."

Accepting these views, of course, and referring to the concluding part of Chapter VII, a further question might be raised at this juncture. Since the alkali metals have atoms of largest diameter (Bragg measurements), this largeness suggests a tendency within the atom to suffer a species of ionisation of its parts—very much as the lithium atoms amongst themselves are virtually ions, even in the dry solid state, according to the views given in Langmuir's paper. It might appear that the energy of the radio-atoms arises from the forces within the atom which are not unlike those which occur in chemical action. In one case helium atoms ($=\alpha$-particles) are hurled out of the main atom with great velocity—considering the most energetic type of radio-action. It is this phenomenon which gives rise to the heat referred to above, or its equivalent in $\alpha$-particle velocity when the particle escapes from the mass. In the other case, the fluorine atoms of the molecule ($F_2$) part company, and take from the lithium atoms, which are ions, their electrons so as completely to denude their outer layers of electrons. The energy involved in the complete process is considerably greater than that given out, since much energy must have been spent in parting the fluorine atoms. The energy in this case seems traceable to the electrons on the lithium atoms, these being in such a state as to leave their atoms, and this is very much like radio-activity,* since electrons are given out radio-actively, if it may be so expressed, in some radio-changes. When, however, the helium atom, which has a mass—

$$\frac{4 \cdot 00}{1 \cdot 008} \times 1845 \dagger = 7320 \text{ times}$$

greater than an electron, is expelled the energy must be this number of times greater, assuming the velocities to be the same. The velocities are different, but the fact remains that the greatest energy is evolved when the $\alpha$-particles are liberated from the atom which is in accord with the dynamic quantities involved. The energy of the $\beta$-particles is comparatively small. H. Herszfinkiel and L. Wertenstein ‡ find that the heat produced by $\beta$-particles from radium B represents much less energy than that from radium C in about the ratio of 2 to 100.

The velocity of $\alpha$-particles from different radio-atoms varies from about 9000 to 13,680 miles per second, light being 186,000 miles per second in a vacuum. Some $\beta$-particles have a speed very near to that of light, e.g. those from radium C have a velocity of 0·995 that of light. Those from radium B are, however, slower, ranging from 0·36 to 0·74 that of light.

Looking at the matter from a more general point of view, each radio-atom in the act of undergoing spontaneous disintegration is like a minute sun in that it gives out enormous quantities of energy, as explained above. Up to the time of spontaneous disintegration the radio-atom remains quite dormant and displays

---

* A difference, however, must be kept in mind, as in radio-changes the action arises from the innermost part of the atom as distinct from chemical action, which involves changes in the external arrangement of electrons.

† This figure, as given by different writers in recent scientific literature, varies from 1836 to 1845 ; see Chapter III.

‡ *Journ. de Physique et le Radium* (1920), 1, p. 143.

no activity and thus, when quiescent, it resembles the atom of ordinary elements.

The question now arises as to whether it will ever be possible to control, or to liberate at will, this atomic energy. So far as present knowledge extends it is somewhat of a mystery.

The sun-energy is also a mystery, as the sun's huge output of radiant energy has not been fully explained. The average reader may not realise the enormous activity of the sun. During the last total eclipse—when Einstein's prediction, that the light propagation in its course from distant stars would be drawn inwards towards the sun when passing this body, was verified—the photographic plates revealed a prominence flame which extended outwards a *hundred thousand miles*.*

The following taken from the *Encyclopædia Britannica* (1911) is of interest: " If we examine the sources for maintenance of the sun's heat, combustion and other forms of combination are out of the question, because no combinations of different elements are known to exist at a temperature of 6000°. A source which seems plausible, . . . is rearrangement of the structure of the element's atoms. An atom is no longer figured as indivisible, it is made up of more or less complex, and more or less permanent systems in internal circulation. Now under the law of attraction according to the inverse square of the distance, or any other inverse power beyond the first, the energy of even a single pair of material points is unlimited, if their possible closeness of approach to one another is unlimited. If the sources of energy within the atom can be drawn upon, and the phenomena of radio-activity leave no doubt about this, there is here an incalculable source of heat which takes cogency out of any other calculation respecting the sources maintaining the sun's radiation." Rutherford † discusses this problem briefly as follows : " It was suggested by Birkeland ‡ in 1896 that the aurora was caused by a type of electric radiation emitted from the sun. Following the discovery of the corpuscular nature of the cathode rays, it was generally supposed that the sun emitted swift $\beta$-particles which penetrated some distance into our atmosphere and gave rise to the luminosity observed in the aurora. The theoretical question of the orbits of the $\beta$-particles in the earth's magnetic field has been worked out and discussed in detail by Störmer.§ Later Vegard ‖ suggested that certain peculiarities occasionally observed in the aurora, for example the curtain or drapery forms, received a simple explanation if it were supposed that the sun emitted radiations of the $\alpha$-ray type. We have

---

\* Prominences may exceed this distance. A recently-recorded prominence shot outwards to a distance of 831,000 kilometres.

† *Radio-active Substances and their Radiations* (1913), p. 654.

‡ *Arch. d. Sci. Geneva* (1896) ; *The Norwegian Aurora Polaris Expedition*, 1902–3.

§ *Arch. d. Sci. Geneva* (1907), 24.

‖ *Phil. Mag.* (1912), 23, p. 211.

seen that the $\alpha$-rays of a given velocity have a definite range in air, and this offers an explanation of the abrupt termination of the luminosity observed in the drapery bands of the aurora. The indirect evidence obtained from the study of the aurora undoubtedly suggests that the sun emits rays similar in type to the $\alpha$- and $\beta$-rays. This would suggest the presence of considerable quantities of radio-active matter near the surface of the sun. So far no evidence has been obtained that a penetrating radiation corresponding to the $\gamma$-rays is emitted from the sun. This, however, does not imply the absence of radio-active matter, for even if the sun were composed of pure radium it would hardly be expected that the $\gamma$-rays emitted would be detectable at the surface of the earth, since the rays would be almost completely absorbed in passing through the atmosphere, which corresponds to a thickness of 76 cms. of mercury.

" From the known data of the absorption of the $\gamma$-rays by air . . . it can readily be shown that the intensity of $\gamma$-rays emitted from the sun at the surface of the earth would only be about $e^{-50}$ of the intensity at the confines of the earth's atmosphere.

" The origin and duration of the sun's heat has been discussed in detail by Lord Kelvin in Appendix E of Thomson and Tait's *Natural Philosophy*. Kelvin calculated the energy lost in the concentration of the sun from a condition of infinite dispersion, and concluded that it seems ' on the whole probable that the sun has not illuminated the earth for 100 million years and almost certain that he has not done so for 500 million years. As for the future we may say, with equal certainty, that inhabitants of the earth cannot continue to enjoy the light and heat essential to their life for many million years longer, unless sources now unknown to us are prepared in the great storehouses of creation.'

" The discovery of the enormous amount of energy emitted during the transformations of radio-active matter renders it possible that this estimate of the age of the sun's heat may be much increased. . . . . Sir George Darwin * drew attention to this probability, and at the same time pointed out that on Kelvin's hypothesis, his estimate of the duration of the sun's heat was probably much too high. He stated that ' The lost energy of the sun, supposed to be a homogeneous sphere of mass $M$ and radius $a$, is $\frac{3}{5}\mu M^2/a$, where $\mu$ is the constant of gravitation. On introducing numerical values for the symbols in this formula, I find the lost energy to be $2 \cdot 7 \times 10^7 M$ calories, where $M$ is expressed in grams. If we adopt Langley's value of the solar constant, this heat suffices to give a supply for 12 million years. Lord Kelvin used Pouillet's value for that constant, but if he had been able to use Langley's, his 100 million would have been reduced to 60 million. The discrepancy between my results of 12 million and his of 60 million is explained by a conjectural augmentation of the lost energy to allow for the concentration of the solar mass towards its central parts.'

* *Nature* (1903), 68, p. 496.

" It appears that the heat emission of the sun cannot be seriously influenced by the presence of known types of radio-active matter, for a simple calculation shows that if the sun consisted entirely of uranium in equilibrium with its products, the generation of heat due to active matter would only be about one-fourth of the total heat lost by radiation.   There is, however, another possible factor which should be considered.   At the enormous temperature of the sun, it appears possible that a process of transformation may take place in ordinary elements analogous to that observed in the well-known radio-elements.   We have seen that the total emission of energy for the complete transformation of one gram of radium is about $3.7 \times 10^9$ calories.   There is every reason to suppose that a similar amount of energy is resident in the atoms of the ordinary elements.   If the atomic energy of the atoms is available, the time during which the sun may continue to emit heat at the present rate may be much longer than the value computed from ordinary dynamical data."

It will be seen that solar physics is of particular interest in con-nection with the atom and its energy.   A brief account of some recent views advanced by M. N. Saha * will be given here under the following title :—

**Elements in the Sun and their Ionisation.**—In studying the chemical elements in the sun, Saha observes that since there is no direct spectroscopic evidence of nitrogen in the sun, nitrogen being a *light* element, and also of thorium, thorium being a *heavy* element, there may be a reason for these and other apparently missing elements not giving spectroscopic evidence of the solar existence. In other words, atomic weight cannot be directly associated with the absence of these elements in the sun's spectrum ;   therefore, some other cause is to be sought.   It is suggested that all the elements are really present in the sun, but only those which are stimulated into activity by the high temperature prevailing therein emit radiation.

It is suggested that the cause of this variation in the sun is due, not to a temperature variation, but to differences in the internal structure of the atoms.

In order to test this view, the elements are classified or studied according to their ionisation potentials.   Thus the most strongly ionised elements appearing in the high-level chromosphere would be Ca, Ba, Sr, Sc, Ti and Fe ; while in the lower layers such atoms are mixed with neutral ones.   Hydrogen ($H_2$) is completely dis-sociated into atoms throughout the solar atmosphere as there is no evidence of a secondary spectrum due to hydrogen molecules.   This would be expected as the temperatures of the chromosphere and photosphere are respectively about 6000° and 7500° abs.

These ideas may now be considered more in detail.   There are a

* *Phil. Mag.* (1920), 40, pp. 472 and 809 : papers A and B.

few quoted passages below which are taken from Saha's paper cited above.

The high-level chromosphere shows lines which, when imitated in the laboratory by using a spark discharge, are stronger than when obtained by means of an arc, and these stronger lines have been styled *enhanced lines*. The elements which give these lines in the solar chromosphere are those metals given above.

The enhancement of the lines in the higher levels of the chromosphere cannot be explained on the theory of a higher temperature in this part of the sun, relative to other parts, as it should be *lower* ; though it would appear to be connected with the physical mechanism of the spark ; and according to Lockyer the spark would represent a greater though more localised temperature as compared with that of the arc. To meet this difficulty, it is suggested that the lines are not due to radiations from the normal atom of the element, but that they arise from an *ionised atom*, i.e. one which has lost an electron. The high-level chromosphere is, according to this view, the seat of very intense ionisation.

Modern theories of the atomic structure and radiation leave little doubt that the enhanced lines in question are due to the ionised atom of the element. " Kossel and Sommerfeld * make it quite clear that the spark-lines of these elements are due to the ionised atom." Lorenser and Fowler † have shown that the series formula for the double lines is of a type which involves $4N$, $N$ being Rydberg's constant, and in the light of Bohr's theory this means that " during the emission of the enhanced lines, the nucleus, and the system of electrons (excluding the vibrating one) taken together behave approximately as a double charge, so that the spectroscopic constant [see Chapter X]$=\dfrac{2\pi^2 e^2 E^2 m}{h^3}$, becomes $4N$, as $E=2e$. This means that the nuclear charge is $n$, the total number of electrons is $(n-1)$, and the system has been produced by the removal of one electron from the normal atom." This observation applies also to the Sr and Ba lines, these elements being ionised.

Considering hydrogen, which is present throughout the atmosphere of the sun, it would appear that there would be little ionisation though complete dissociation of this element has taken place ; and the greater ionisation potential of this element under a given thermal stimulus, as well as that of helium, shows that these elements cannot undergo ionisation anywhere in the sun to an appreciable extent.

Helium is more ionised in the hotter stars where the temperature is in the neighbourhood of 16,000° abs., and these stars show the Rydberg line 4686 and the Pickering lines.

In order to study more closely this view of solar radiation it is necessary to compare the ionisation potentials (see Chapter XII)

---

* *Ber. Deutsch. Phys. Gesell.* (1919), 21, p. 240.
† See Saha's papers for further particulars.

of the elements which appear to be absent from the sun, as well as those present.   To this end, the ionisation of an atom $A$ takes place according to the chemical scheme $A \rightleftarrows A_+ + e - U$, in which $U$ is the quantity of energy liberated in the process.   This equation or scheme of action is based upon Nernst's theorem for the " *Reaction-isobar* " by assuming that ionisation is a sort of reversible chemical process.   $U$ is deducible from the ionisation potentials (V) according to the energy equation $\dfrac{eV}{300} = h\nu_o$, $\nu_o$ being the convergence frequency of the principal series, using Paschen's notation with Sommerfeld's modification so that $1 \cdot 5s$ becomes $1s$.   Multiplying the quantity thus obtained by Avogadro's number (this constant may lie somewhere between $6 \cdot 07$ and $6 \cdot 90 \times 10^{23}$—a mean value being $6 \cdot 5 \times 10^{23}$ ; but according to Millikan's measurements the lowest value seems most probable), the energy $U$ may be obtained in calories. Taking a single example, magnesium has an ionisation potential of $7 \cdot 65$ volts, and the calculated calories by the above equation comes out at $1 \cdot 761 \times 10^5$ (ergs being converted into calories).

Having calculated the energy necessary to ionise the elements, it will be seen that the almost complete ionisation of Ca, Sr and Ba atoms in the high-level chromosphere is due to the low pressure in these regions.   " Moreover, the greater the ionisation potential of an element the more difficult ionisation will be for that element under a given thermal stimulus."

The study so far recorded is really a theory of temperature-pressure ionisation as applied to the sun.   Summing up : " Elements which are missing from the sun can be broadly subdivided into two groups—

> " 1st, those which are completely ionised, e.g. Rb, Cs and probably Tl [see Table XXI in next section].
>
> " 2nd, elements of which the ionisation and radiation potentials are so high that they are not in a state capable of absorbing those of their characteristic lines which occur in the continuous photospheric spectrum (*vide* paper C, *Phil. Mag.* (1921), 41, p. 267).

" Helium and most of the inert gases fall within this group.   Neon, argon and other inert gases have very high ionisation potentials, and their presence cannot be detected, for it is shown that the varying records of different elements in the Fraunhofer spectrum may be regarded as arising from the varying response of these elements with regard to the stimulus existing in the sun.   The stimulus existing in the sun is the same for all elements, namely, that arising from a temperature of about $7500°$ K., but owing to different internal structure, elements will respond in a varying degree to this stimulus" (see Table XXI, columns $a$, $b$, $c$, $d$ and $g$).

In a recent contribution by Rutherford and Chadwick * it is stated that the energy from the H-atoms liberated from aluminium is 25 per cent greater than that of the incident α-particle, thus showing that it is possible to set free some of the internal energy of the atom. It had previously been shown that in bombarding nitrogen with α-particles the energy set free possibly amounted to about 8 per cent of the energy of the incident α-particle. The energy of the atom will be further discussed in the light of these experiments of Rutherford and the data given in Saha's paper (B) under the following title :—

**Atomic Energy and Solar Radiation.†**—The enormous output of radiant energy from the sun has been attributed to two causes, viz. (1) contraction of the sun's mass, and (2) radio-active disintegration of its atoms. It may appear from the following

### TABLE XX

| | |
|---|---:|
| Mass of the sun taking the earth as a unit . . . | 329,390 |
| Mean mass of radio-atoms (H=1·00 or 1·008) . . | 222 |
| Mean mass of radio-atoms taking the mass of the electron as unity, about . . . . . . | 404,000 |
| Mass of the hydrogen atom taking the electron as a unit | 1,840 |
| Temperature of the sun's photosphere, about . . | 7,500° abs. |
| Temperature of the sun's chromosphere, about . | 6,000° abs. |
| Temperature of the electric arc, maximum, about . | 4,200° abs. |
| Temperature of the electric arc, minimum, about . . | 3,750° abs. |
| Instantaneous temperature of exploded wires (see below), about . . . . . . . . | 20,000° abs. |
| The temperature of the sun can be evaluated from its total radiation on the assumption that it radiates as a black body and knowing its *solar constant* which may be taken as 2·5 calories and making use of Stefan's law . . . . . . . . | = 6,200° abs. |
| Estimated pressure at the centre of the sun = about 80 billion pounds per square inch. | |

that views are advanced by the writer which are partly in direct conflict with those of Saha. The answer to this suggested criticism is that it is the object of the writer to present as many views as possible so long as there is some agreement with experiment in each case. Moreover, it seems probable that a certain amount of truth is contained in each view expressed, but not the whole truth.

Proceeding with the argument, it is suggested that there may be a connection between (1) and (2) above. It is of course known that the heavier atoms of the Periodic Table exhibit radio-active properties ; and the mass of the sun is so great (see Table XX) that its contraction would account for the emission of considerable radiant energy ; but, in the latter case this energy has not been

---

* *Nature* (1921), 107, p. 41 : abstract given at end of Chapter VIII.
† This section and the one following it were published in nearly the same form in the *Chemical News* of 13 May and 17 June 1921.

supposed to be sufficient to account for the enormous output of energy known to take place.

These matters may be studied from a general point of view, and it would seem that in some way *pressure due to largeness of mass* might be a factor in both phenomena.

If one considers the atomic edifice to have a fixity of structure and a cohesiveness * which are due to the interaction of positive nuclei and revolving electrons, then it seems only natural that the atoms should have atomic domains (Bragg). The rigidity of matter may be therefore traced to the electromagnetic system of the atom and, if this is encroached upon or sufficiently disturbed, an outburst of energy in the form of ejected particles may take place. In short, all atoms would be radio-active under suitably impressed conditions. If these ejected particles strike into the atoms or molecules of a gas, ionisation results ; and where there is ionisation of an intense kind there is intense radiation.

In the sun it is conceivable that the contraction of its more central mass forces the atomic constituents closer together so that radio-activity is induced ; and, of course, following from this there would be gaseous ionisation and radiation.

Similarly, in the case of the radio-elements composed of more massive atoms, there may be an unstable state set up by the very massiveness of the atom and under just the right combination of conditions (whatever these may be) the atom partly breaks down ; or, in more technical terms, radio-active disintegration takes place.

Therefore, the cause of radiant-energy emission from the sun may be due primarily to the pressure impressed upon its more central mass by contraction, and the atoms suffering such a pressure, perhaps combined with high temperature, may break up or disintegrate, and the internal energy of the atoms liberated as in radio-active phenomena. In fact, one could imagine the sun to be, as it were, a huge radio-active atom whose mass is so great that the forces of its ultimate parts become disturbed ; and a radioactive process the same as that of a disintegrating atom of radium, for example, takes place, thus giving effect to an enormous output of radiant energy. The solar envelope may be bombarded from within by nuclear particles and electrons from disintegrating atoms, which would cause gaseous ionisation.

If this idea has any basis of truth in it, it should be possible to induce radio-activity by subjecting atoms to enormous pressure. To produce artificially sufficient general pressure to cause atoms to disintegrate may not be physically possible, yet it is conceivable, and that is perhaps a step. Sir E. Rutherford's experiments, referred to above, seem to be in harmony with this idea, for Rutherford has succeeded in breaking up the nitrogen atom by bombarding it with high-speed α-particles. In the case of aluminium, the bombardment gave rise to the liberation of H-atoms which have an

* A term borrowed from physics, but applied to the atomic structure.

energy 25 per cent greater than that of the incident $\alpha$-particles. This experiment seems highly suggestive in this connection, for here is a case of an ordinary atom being artificially disintegrated by the blow or instantaneous pressure of the $\alpha$-particles from radium C.

It might, therefore, be possible to trace a connection between the spectrum lines of the elements in the sun and the susceptibility of such elements to disintegration by the method of Rutherford

TABLE XXI

| (a) | (b) | (c) | (d) | (e) | (f) | (g) |
|---|---|---|---|---|---|---|
| Elements not found in the sun | Elements of doubtful existence in the sun | Elements giving faint lines in the Fraunhofer spectrum | Elements which appear not to have been investigated as regards their presence in the sun | Elements whose atoms give long-range H-atoms under bombardment with $\alpha$-particles from radium C | Elements whose atoms do not appear to be disintegrated by $\alpha$-particle bombardment; or which do not give long-range H-atoms on bombardment | Ionised atoms of the high-level chromosphere; and other atoms (elements) known to be present in the sun: but the following list may be extended |
| B | Ne | K | F | B | He | Ca |
| N | Ar | Cu | Cl | N | Li | Sc |
| P | Kr | Zn | Br | F | Be | Ti |
| S | Xe | Ge | I | Na | (C) | Fe |
| As | Ru | Ag | Te | Al | (O) | Sr |
| Se | Ta | Cd | etc. | P | Mg | Ba |
| Rb | W | Sn | | | Si | |
| Sb | Os | Pb | | | S | H |
| Cs | Ir | | | | Cl | He |
| Pr | Pt | | | | K | O |
| Tl | Ra | | | | Ca | Na |
| Bi | Th | | | | Ti | Mg |
| | U | | | | Mn | |
| | | | | | Fe | |
| | | | | | Cu | |
| | | | | | Sn | |
| | | | | | Au | |

Probably Li is present in the sun. See A. S. King, *Astrophys. Journ.* (1916), 44, p. 169. O and C appear to be unstable: see pp. 46, 51.

just referred to. The absence, for instance, of certain lines in the sun's spectrum might be due to the chemical elements having been already broken up by the radio-change induced by the pressure, thereby leaving the more or less unbreakable ones in greater evidence. This observation is supported by the absence of nitrogen, boron and phosphorus in the sun's spectrum, as these elements are more easily broken up by bombardment with $\alpha$-particles than many others which are present in the sun's spectrum. Table XXI gives at a glance the whole position of the elements in question and those which Rutherford has, and has not, been able to disintegrate. It will be seen that those elements the atoms of which resist disintegration are more abundant in the sun, with few exceptions; but the nature of the element in the sun may further modify the conditions, as indicated by Saha. See Note on page 172.

**Sun-Spots, Aurora and Magnetic Storms.**—From the foregoing it will be seen that the phenomena of sun-spots, aurora and magnetic storms are intimately connected evidently with the suggested process of radio-change or bombardment taking place within the sun. It will not be out of order, therefore, to discuss the subject here.

The origin of the aurora in the polar regions where the magnetic pole plays an important part (see above) in giving rise to various electrical effects, as seen in the vacuum tube, has led to certain well-known investigations of this phenomenon. Some of these experiments will be briefly noted below; but, first, an early account of the aurora itself, as given by Humboldt, will serve to remind the reader of the complexity and beauty of the phenomenon. " An Aurora Borealis [*Northern Light*, or *Polar Light*; there is also an *Aurora Australis*] is always preceded by the formation in the horizon of a sort of nebular veil, which slowly ascends to a height of 4°, 6°, 8° and even 10°. It is towards the magnetic meridian of the place that the sky, at first pure, commences to get brownish. Through this obscure segment, the colour of which passes from brown to violet, the stars are seen as through a thick fog. A wider arc, but one of brilliant light, at first white, then yellow, bounds the dark segment. Sometimes the luminous arc appears agitated for entire hours by a sort of effervescence and by a continual change of form before the rising of the rays and columns of light which ascend as far as to the zenith. The more intense the emission of the polar light the more vivid are its colours, which from violet and bluish-white pass through all the intermediate shades to green and purple-red. Sometimes the columns of light appear to come out of the brilliant arc mingled with blackish rays similar to a thick smoke. Sometimes they rise simultaneously in different points of the horizon; they unite themselves into a sea of flames, the magnificence of which no painting could express, and at each instant rapid undulations cause their form and brilliancy to vary. Motion appears to increase the visibility of the phenomenon. Around the point in the heaven which corresponds to the direction of the dipping needle, the rays appear to assemble together and form a boreal corona. It is rare that the appearance is so complete and is prolonged to the formation of the corona, but when the latter appears it always announces the end of the phenomenon. The rays then become more rare, shorter, and less vividly coloured. Shortly nothing further is seen on the celestial vault than wide, motionless, nebulous spots; they have already disappeared when the traces of the dark segment, whence the appearance originated, are still remaining on the horizon." *

De La Rive was probably the first one to suggest that the aurora was an electrical effect as exhibited in Geissler's tubes especially under the modifying influence of a magnet.

Sabine showed that there are periods of maximum frequency of magnetic storms and aurora which coincide with the maximum appearances of sun-spots.

Birkeland devised experiments by which he demonstrated that the magnetic storms could be imitated by placing a spherical body with magnetic poles—termed a *terrella*—inside a vacuum tube, the cathode representing the sun. For example, during an equatorial storm when the horizontal magnetic component is decreased, the conditions can be demonstrated by means of the terrella, which corresponds to the earth. In this case the cathode rays encircle the magnetic equator.

Störmer, in co-operation with Birkeland, worked out the trajectories of the electrified particles on entering the earth's field, the results of which support Birkeland's theory that the phenomenon arises from charged particles ejected from the sun.

Villard in 1906 showed that cathode rays, as influenced by a magnetic field, give rise to effects identifiable with or simulating those of the aurora. D. Owen gives in his book † an account of some of the above experiments ; but see the quotation from Rutherford's book above—especially in reference to the curtain effects showing an abruptness of termination in the luminosity suggesting the limiting range of the particles entering the terrestrial atmosphere.

V. M. Slipher ‡ gives evidence of the constancy of the phenomenon, indicating that normally the sun emits (according to the views here presented) a sufficient number of electrified particles to cause weak effects—in fact, in polar regions the aurora is a regular occurrence. Slipher states that during a period of three and a half years a hundred spectrograms were taken at the Lowell Observatory of the night sky, and each one gave the chief auroral line. The wave-length of this line was found to be 5578·05. It had been thought that this line was due to the nitrogen pair $\lambda\lambda 5560$–5565,§ but this supposition seems no longer tenable.

Passing now to sun-spot phenomena, the following from the *Encyclopædia Britannica* will serve to introduce the subject: " Each spot shows, as a rule, one or more spots or groups of spots. Each spot shows with more or less completeness a ring-shaped penumbra enclosing a darker umbra ; the umbra, which looks black beside the photosphere, is actually about as brilliant as lime-

---

* The reader will find detailed accounts of much later date in the *Encyclopædia Britannica*.
† *Recent Physical Research* (1913) ; see Chapter VI.
‡ *Astrophys. Journ.* (1919), 49, p. 266.
§ See Stark, *Ann. d. Physik* (1918), 54, p. 598.

light." This appearance can be roughly illustrated by dropping a drop of blue-black ink on a piece of paper; and after blotting, dropping a small drop of Indian ink on top of the spot so as to leave a blue-black margin. This margin, of course, represents the penumbra.

A. L. Cortie,* referring to the band spectra of sun-spots and their identification as flutings, remarks on the existence of chemical compounds in these areas. For bands at the extreme red have been identified by Hale and Adams † as those produced by titanium oxide, and later Fowler ‡ has ascribed the bands in the green part of the spectrum to magnesium hydride. The observed change in the spectrum of Mira Ceti—as the brightness and temperature of

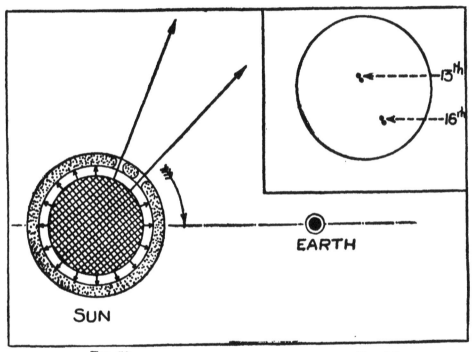

FIG. 56a.          FIG. 56b.

this star had increased, its banded spectrum had gradually changed more to a line one—affords evidence that the band spectra of the sun-spots indicate a temperature of the vapour therein which is lower than that of the photosphere.

Referring to Figs. 56a and b, the upper figure represents the spots as they appeared during the interval May 13–16, 1921, when seen through an opera-glass, the sun being low and slightly dimmed by haze. The change in position seems to agree with the sun's rotation (25 days 9 hours 7 minutes), but no attempt has been made to give an accurate drawing. Accompanying the recent

* *Astrophys. Journ.* (1907), 26, p. 123.
† See W. S. Adams, *ibid.* (1909), 30, p. 86.
‡ *The Observatory* (1907), 30, p. 272.

auroral displays there was a magnetic storm of unusual violence. These were recorded in the press at the time. Referring to the *a*-diagram (not drawn to any particular scale and not intended to conform to any detailed features of the sun's structure, some of which are superficially known), it will be seen that as the spots revolve the ejected matter, according to the views given above, represented by the long arrows, will pass within range of the earth and come under the influence of its magnetic field, so that, assuming the ejected matter to be charged particles, when they strike into the earth's atmosphere they will give rise to the aurora and to the accompanying magnetic storm.

The blackness of the spots perhaps has a partial analogy in the Faraday and Crookes dark spaces as seen in vacuum tubes, especially as according to the views here advanced the whole phenomenon is a bombarding process involving positive-rays, or atomic-nuclei and probably negative electrons. The radial arrows shown in the figure are intended to illustrate the bombardment or the general course of the corpuscular rays.

There seems to be some evidence that the sun-spots are due to a species of eruption which has its origin in the nucleus of the sun. A gas composed of the above-mentioned particles might accumulate in the central part or core of the sun and work out gradually to the surface. The sun-spot period of maximum activity is eleven years, which corresponds to about 158 revolutions of this body on its axis.

A thinning of the envelope may take place at times or in areas, and considerable quantities of the particles of disintegration may pass completely away from the sun. This would perhaps account for magnetic storms when the spots are not in particular evidence, and certainly it would seem to account for the auroral line always being present in the night sky.

There are peculiarities about the spots themselves showing evidences of powerful cyclonic disturbances, but the entire surface of the sun is in violent agitation and the prominence flames extend outwards thousands of miles.* The whole phenomenon is on such a gigantic scale that one can hardly realise its magnitude. The spots themselves are often thousands of miles across. In the recent display, cited above, their widest part was reported to be about 80 or 90 thousand miles.

The reader should bear in mind that any views advanced which are not yet supported by reliable and repeated experiments are subject to modification when new evidence becomes available.

*Summarising the foregoing*, it was suggested that atomic nuclei and electrons were ejected from the more central part or nucleus of the sun owing to a pressure effect therein. It is now further suggested that some of these particles from the nucleus escape more or less radially through rifts in the solar envelope (which are the sun-spots) and when these particles enter the earth's atmosphere an aurora

* A very high flame would perhaps reach a sun altitude of 500,000 miles.

and magnetic storms are produced in accordance with the ideas of Birkeland and others. Magnetic fields have to be taken into account in tracing the direction of the particles.

**High-Pressure Experiments of Sir C. A. Parsons.**—It is of interest to note here some recent experiments of Parsons * involving very high pressures. In attempting to produce diamond, pressures were obtained reaching 15,000 atmospheres per square inch and an instantaneous pressure twenty times as great was theoretically obtained, which, it was stated, was about one-half that at the centre of the earth, while in many stars the pressure would be enormously greater ; the above value would be " only a fraction of one-thousandth part the pressure at the centre of large stars." † In one experiment in which a steel piston was driven by the explosion of powder against a gas charge of oxygen and acetylene contained in a bored steel block, the pressure was supposed to be about 15,000 atmospheres per square inch and the calculated instantaneous temperature when explosion of the gas took place reached the enormous figure of 17,700° C. Further experiments appear to be in progress.‡

From certain mechanical or dynamic considerations perhaps it is not correct to consider such enormous pressures, as suggested in the case of the disintegration of atoms in the sun, to act as a displacement blow or shear, and consequently *spin energy*, as it might be termed, may become a factor. One has to consider the structure of the nucleus of the atom in this connection. Then in turn, centrifugal force would have to be taken into account. However this may be, besides this latter consideration the favourable chance positions of the atoms or their nuclei possibly enters into the problem.

**High Temperatures Obtained by Exploding Wires Electrically.** —It is important to investigate stellar phenomena in the laboratory, as thereby a better understanding of stellar physics is obtained. To this end, J. A. Anderson § has investigated the light from electrically-exploded wires with the object of imitating high-temperature absorption spectra. When this method is fully developed it is hoped to reproduce stellar absorption spectra of the solar type. The experiments are, moreover, of interest in showing the possibility of obtaining very high instantaneous temperatures ; for in probing the atom it is important to consider high temperatures and

---

* *Inst. of Metals Journ.* (1918), 20, p. 5.
† The pressure at the centre of the sun is enormous. Taking that of one pound on a square inch at the earth's surface as a unit, the pressure at the sun's centre would be about 80 billion times greater, or 80 billion pounds per square inch.
‡ See reports of *Roy. Soc. Proc.* in *Chem. News* (1921), 122, p. 235.
§ *Mt. Wilson Observatory Contributions*, No. 178 ; or *Astrophys. Journ.* (1920), 51, p. 37.

any effects arising therefrom in connection with atomic structure—
in particular the study of the problem from the energy side.

In these experiments a large condenser charged to 26,000 volts
was discharged through a fine wire (5 cms. long, weight 2 mgrms.)
in about $10^{-5}$ second. It was shown that about 30 calories were
dissipated during this short interval of current flow. The wire was
volatilised in a confined space. If all the energy involved had been
dissipated as heat in the wire itself its temperature should have
been about 300,000° C. When the wire exploded the flash intensity,
however, corresponding to a temperature of about 20,000° C.,
was about one hundred times greater in brilliancy than that of
the sun.

When the wire was exploded within a tube or slot under a bell jar
exhausted to 20 mm. pressure a line emission spectrum was obtained, and as the pressure was increased the continuous background increased in intensity and the spectra became more an
absorption one. The spectrograms showed clearly a pressure
shift of the lines. Pressures up to 50 atmospheres can easily be
obtained.

Using an iron wire, an absorption spectrum was produced which
extended to 0·66 $\mu$,* and this spectrum includes all classes of lines
except pronounced enhanced lines. It is hoped to reproduce
stellar absorption spectra of the solar type by this method.

* Absorption spectra obtained with the arc in the electric furnace had been
limited to wave-lengths below 0·56 $\mu$ for iron, and high-temperature lines had not
been recorded.

## NOTE

With regard to the non-existence of certain elements in the
sun, it should be borne in mind that some types of atoms may be
partly disintegrated in the sun, as above suggested, and the
question then arises as to the spectrum emitted by such atoms.
If, for example, nitrogen has been partly broken up by solar
bombardment, after the manner of Rutherford's experiment (see
Chapter VIII), would its spectrum suffer any change in consequence? This consideration, as well as the other views stated
in this chapter, including those of Saha, should be kept in mind.
One must keep in view a multiplicity of ideas until experiment
reveals the true state of affairs.

# CHAPTER XXII

## ATOMIC AND MOLECULAR MAGNITUDES DETERMINED BY THE BROWNIAN MOVEMENT

PROF. JEAN PERRIN delivered a discourse on the Brownian movement and 'size' of molecules before the Royal Institution in 1911,
and since then (in 1916) a work by Perrin * has appeared.† It
will only be necessary to touch briefly on this interesting subject,
as full particulars are given elsewhere, as indicated.

The Brownian movement consists of to and fro excursions of
small particles, the movement being initiated by molecular movements. This movement need never cease, as it has been observed
in liquids imprisoned in quartz for thousands of years. The
movement is general, only it escapes observation without special
magnifying apparatus. For example, the particles forming tobacco
smoke are continuously kept in motion by the moving molecules
of the air quite apart from air currents.

Camphor particles sprinkled on water serve to illustrate the
type of movement to the unaided eye. They appear as if they were
constantly being struck by the water molecules which are not
quiescent. As a matter of fact, the movement of the camphor
particles is due to an irregular dissolving of the camphor which sets
up one-sided surface tensions that pull the particles about on the
surface of the water.

All bodies have natural periods of vibration, consequently they
are more or less susceptible of movement, and when they are very
small the movement is quicker. Large particles, like those of
camphor, cannot be visibly moved by the molecular impacts, nor
can they themselves partake of any visible Brownian movement.

These particles obey mechanical laws and, when the theory
of gases is extended to their movements, the experimental facts
fit in with the theoretical deductions. For example, in a vertical
column of air the distribution of the gas-molecules obeys Boyle's
law (see Chapter I), to the effect that the density of the air in going
upwards decreases in geometrical progression—that is to say,
taking a given distance, 6 kilometres in this case, the density

---

* *Atoms* (1916). See also by same author, *Brownian Movement and Molecular
Reality* (1910) : translated by Soddy.

† See also report of Perrin's lecture in *Chem. News* (1912), 106, pp. 189, 203
and 215.

diminishes one-half at every six-kilometre step upwards, starting either at sea-level or at a very high altitude.

Now, Perrin found that the small particles studied in the Brownian movement underwent the same distribution, thus obeying the above law. Emulsions were studied, and under the ultra-microscope countings were made, and finally the laws verified, so that it is only a step in the chain of connected magnitudes to evaluate from the action of visible particles the mass of the hydrogen atom itself, which comes out at $1.47 \times 10^{-24}$ gramme. The kinetic theory of gases leads to the value $1.6 \times 10^{-24}$. The agreement is therefore good.

Einstein pointed out that knowing the linear displacement of a given particle in a given time would enable the mass of the particle to be determined. No matter how many zigzag excursions a particle may take in going from place A to place B the direct distance from A to B is taken (in a given time) as the displacement. A number of measurements have to be made so as to get a true *mean* figure in applying Einstein's law. In practice the particles under study were photographed at intervals by means of a cinematograph. By this method the reasoning could again be extended to hydrogen and its mass deduced, which came out at $1.4 \times 10^{-24}$ gramme.

Perrin says that the discontinuity of matter is shown by the fact that there is convergence, not only in each method, but in all methods. These involve phenomena which are as fundamentally different as viscosity of gases, the Brownian movement, the blueness of the sky, the electrification of droplets, radio-activity, and distribution of energy in the spectrum. "These facts demonstrate the objective reality of the molecules and give the same magnitude for the elements of matter."

# CHAPTER XXIII

## MAGNETIC SUSCEPTIBILITIES OF THE ELEMENTS*: A GEOMETRICAL REPRESENTATION OF VARIABLE MAGNITUDES: FERROMAGNETISM AND ATOMIC STRUCTURE

THE use in the text of such terms as magnetic susceptibility, permeability and lines of magnetic force calls for some explanation, as some readers may not be familiar with the meaning of these terms. Moreover, the magnetic properties of the elements become those of the atom and molecule, and in studying atomic structure magnetism may come more into prominence as atomic phenomena are further investigated. Therefore, a short account of magnetic susceptibility along general lines will be introduced here. In using the expression *lines of magnetic force*, it will be shortened to *magnetic lines* and in some cases the word *lines* only will be used.

The well-known expression—

$$\mu = \mathbf{B}/\mathbf{H} \qquad \qquad (1)$$

indicates that $\mu$ is the ratio derived from the number of magnetic lines, $\mathbf{B}$, per unit cross-section (square inch or square centimetre) in iron, for example, divided by $\mathbf{H}$, which is the number of lines that would be developed in the same unit area of air or empty space, the magnetising force being the same in each case. That is to say, a coil carrying a given current will produce more lines in iron (situate within the coil) than in the air when the iron is removed from the coil, and this ratio, whatever it may be in a given case, is termed the *permeability* ($\mu$).

The permeability varies with the quality, state, history, magnetisation, etc., of the iron, and whether it is soft, hard, etc. The magnetising force, $\mathbf{H}$, coincides with the magnetic lines producible in the air or empty space, and these descriptive terms are exchangeable, as space or air free from magnetic substance has no varying permeability or multiplying power, being taken as *unity*. In other words, the number of magnetic lines in free space or air numerically represents the magnetising force. The expression " magnetic lines of force " is supposed by some to have a physical meaning, as if the

---

* Reproduced mainly from an article by the writer in the *Chemical News* (1914). Kind permission has been given by the editor, J. H. Gardiner, to make use of any articles by the writer in this publication.

magnetism actually extended in thread-like lines. On the other hand, others regard this term as too descriptive, and they say that it should only be accepted as conventional, and that no such *line* conception should be entertained. However, the picturesque term is preferable, but the possibility, or even probability, that it is merely a conventional device should be kept in mind. It is true that this conception implies *direction* and the number of lines indicate *strength*, so that the expression is quite in order.

From Table XXII these variables may be noted. This table is copied from *C.G.S. System of Units* (1891), p. 148, by J. D. Everett, but with some values omitted, the figures being based upon the early

### TABLE XXII

| H<br>Magnetising force | I<br>Magnetisation | k<br>Susceptibility | B<br>Induction | μ<br>Permeability |
|:---:|:---:|:---:|:---:|:---:|
| 0·3 | 3 | 10 | 41 | 128 |
| 2·2 | 117 | 53 | 1460 | 670 |
| 4·9 | 917 | 187 | 11540 | 2350 |
| 6·7 | 1078 | 161 | 13520 | 2020 |
| 10·2 | 1173 | 115 | 14840 | 1450 |
| 78·0 | 1337 | 17 | 16900 | 215 |
| 208·0 | 1452 | 7 | 18500 | 89 |
| 585·0 | 1530 | 2·6 | 19800 | 34 |
| 24500·0 | 1660 | 0·067 | 45300 | 1·9 |

measurements of Ewing and Bidwell on iron, which may be taken as fairly characteristic.

In this table the term $k$ is given, which is the *magnetic susceptibility*. Iron, cobalt and nickel have a high susceptibility which passes through a wide range of magnitude as compared with all other elements (see Table XXIII), $k$ being particularly variable in highly magnetic (ferromagnetic) metals, such as iron. The susceptibility is also a ratio, like the permeability, being the ratio of the intensity of the *induced* magnetism (I) to the magnetising force (H) ; the relation may be expressed thus—

$$I/H = k \quad . \quad . \quad . \quad . \quad . \quad . \quad (2).$$

The term **I** at first glance seems unnecessary, as it is already seen that the magnetising force (**H**) gives rise to a given number of lines per square cm. (**B**) in the metal or substance, according to its permeability ($\mu$). By a somewhat artificial procedure, the induction

B may, however, be divided into two components, *one* being the magnetic lines due to the magnetising force **H**, as produced in empty space or air, and the *other* due to the magnetic lines which are *induced* or called into existence, or influenced, so to speak, by the peculiarity of the substance acted upon by the magnetising force ; whether, for example, it be *paramagnetic* like iron, or *diamagnetic* like bismuth ; therefore, calling the intensity of magnetisation due to, or induced through, the agency of the substance **I**, and the lines that would be produced in air or empty space **H**, which becomes the magnetising force, the following equation represents the union of these two terms :—

$$\mathbf{B}=4\pi\mathbf{I}+\mathbf{H} \quad . \quad . \quad . \quad . \quad . \quad . \quad (3).$$

This equation is truly *algebraic*, since it holds good if the signs change ; see below. The $4\pi$ factor, or multiplier, arises from the old conception of $4\pi$ lines emanating from a sphere of unit radius (=unit pole), and appears ⠂ the calculations of the magnetic circuit.

It is therefore apparent that if the substances exercise variable positive and negative modifying influences, the magnetic susceptibility of any particular substance may be defined by the ratio of **H** to **I**, as shown by equation (2), and this ratio, moreover, may be positive or negative, according to the inherent nature or characteristic of the material (see Tables XXIII and XXIV).

Having established the fundamental relations, it is obvious that different formulæ may be constructed thus :—

It was shown that—

$$\mathbf{B}=4\pi\mathbf{I}+\mathbf{H},$$

and by definition of $k$—

$$\mathbf{I}=k\mathbf{H} ;$$

hence—

$$\mathbf{B}=4\pi k\mathbf{H}+\mathbf{H}=(4\pi k+1)\mathbf{H},$$

whilst—

$$\mathbf{B}=\mu\mathbf{H} ;$$

therefore—

$$\mu=4\pi k+1$$

and—

$$k=\frac{\mu-1}{4\pi} \quad . \quad . \quad . \quad . \quad . \quad . \quad . \quad . \quad (4),$$

which are all well-known expressions (see Ewing, *Magnetic Induction in Iron and other Metals*, 3rd edition (1910), pp. 12 and 18).

While the permeability of air is taken as 1, and its susceptibility as zero, when studying the permeability and susceptibility of highly magnetic substances such as iron, cobalt, etc.; *air*, as a matter of fact, is very feebly magnetic, and for an absolute comparison (or correction) a vacuum may be taken, which is absolute unity in one case and absolute zero in the other.

The value of $k$ for water at 20° C. $= -0.75 \times 10^{-6}$.*

Referring to Table XXIV, it will be seen that no element has been discovered which has a negative susceptibility sufficiently

TABLE XXIII.—*Paramagnetic Elements*

| | Positive susceptibilities for unit mass, given in $10^6 \times k_m$ values; $k_m$ being the susceptibility, which may be calculated by multiplying these values by $10^{-6}$ | Corresponding values for unit volume, $10^6 \times k_v$ |
|---|---|---|
| Iron . . . | $+31,170,000$ (a) wrought iron | |
| Cobalt . . | $+ 1,842,000$ (b) cast cobalt | |
| Nickel . . | $+ 2,003,000$ (c) annealed nickel rod | |
| Oxygen . . | | $+324$ at $-182°$ |
| Manganese . | $+20$     at 1000° ($+11$ at 18°) | |
| Palladium . | $+5.8$   ,,    18° ($+2$ at 1100°) | $+50$ to $+60$ |
| Chromium . | $+4.2$   ,, 1100° ($+3.7$ at 18°) | |
| Titanium . | $+3.5$   ,, 1100° ($+3.1$ at 18°) | |
| Rhodium . | $+1.9$   ,, 1150° ($+1.1$ at 18°) | $+13$ |
| Vanadium . | $+1.8$   ,, 1100° ($+1.5$ at 18°) | |
| Niobium . . | $+1.3$   ,,    18° | |
| Platinum. . | $+1.1$   ,,    18° ($+0.7$ at 1000°) | $+29$ |
| Tantalum . | $+0.93$ ,,    18° ($+0.8$ at 800°) | |
| Beryllium . | $+0.79$ ,,    15° | |
| Aluminium . | $+0.65$ ,,    18° ($+0.5$ at 1000°) | $+1.7$ to $+1.9$ |
| Magnesium . | $+0.55$ ,,    18° | |
| Sodium . . | $+0.51$ ,,    18° | |
| Potassium . | $+0.40$ ,, 18—180° | |
| Lithium . . | $+0.38$ | |
| Tungsten . | $+0.33$ ,, 18—1100° | |
| Iridium . . | $+0.3$   ,, 1100° ($+0.15$ at 18°) | $+4.9$ |
| Thorium . . | $+0.3$   ,,  400° ($+0.18$ at 18°) | |
| Osmium . . | $+0.04$ ,, 18—1100° | |
| Molybdenum | $+0.04$ ,,    18° | |
| Tin. . . . | $+0.03$ ,, 18—240° | $+0.35$ |
| Nitrogen at atmospheric pressure. . | | $+0.001$ at 16° |

EXPLANATORY NOTE.—The second column gives the values ($10^6 \times k_m$) for elementary substances of the first column (not single atoms). Values marked a, b, c are fairly characteristic, and show the enormously greater susceptibilities of these elementary substances under conditions of practically the highest susceptibility. The bracketed values show lower values for the temperatures (° C.) specified. Some of the variations due to temperature are no doubt due to experimental errors, or they arise in some cases from iron as an impurity. In Honda's experiments, the effects of iron were either eliminated at the outset, or when purity could not be obtained corrections for traces of iron being present were made.

high to make the permeability ($4\pi k + 1$) negative. A *paramagnetic* substance has a permeability greater than 1, and a positive sus-

* H. C. Hayes (1914) obtains the value $-0.726 \times 10^{-6}$ at 24° C., which agrees well with the determinations of Sève, de Hass and Drapier, and Weiss and Picard.

ceptibility. A *diamagnetic* substance has a permeability less than unity or empty space, and a negative susceptibility. No known substance is more than slightly diamagnetic, bismuth having a negative volume susceptibility of $-16 \times 10^{-6}$, and a corresponding permeability only slightly below 1, namely, $0 \cdot 9998$.

Tables XXIII and XXIV are self-explanatory. The values, except *a*, *b* and *c*, are from the fifth edition of *Landolt-Börnstein*

TABLE XXIV.—*Diamagnetic Elements*

| | Negative susceptibilities for unit mass, given in $10^6 \times k_m$ values; $k_m$ being the susceptibility, which can be calculated by multiplying these values by $10^{-6}$ | Corresponding values for unit volume, $10^6 \times k_v$ |
|---|---|---|
| Copper . . . | $-0 \cdot 09$ at 18—1000° | $-0 \cdot 66$ to $-0 \cdot 80$ |
| Silicon . . . | $-0 \cdot 12$ ,, 18° | |
| Lead . . . . | $-0 \cdot 12$ ,, 18—330° | $-1 \cdot 4$ to $-0 \cdot 84$ |
| Zinc . . . . | $-0 \cdot 15$ ,, 18 ($-0 \cdot 10$ at 650°) | $-0 \cdot 70$ to $-1 \cdot 0$ |
| Gold . . . . | $-0 \cdot 15$ ,, 18—1060° | $-3 \cdot 1$ |
| Cadmium . . | $-0 \cdot 17$ ,, 18° ($-0 \cdot 15$ at 700°) | |
| Mercury. . . | $-0 \cdot 19$ ,, 18—250° | $-2 \cdot 6$ at 19° |
| Silver . . . | $-0 \cdot 22$ ,, 1100° ($-0 \cdot 19$ at 18°) | $-1 \cdot 7$ at 15° |
| Arsenic . . . | $-0 \cdot 3$ ,, 18—200° | |
| Selenium . . | $-0 \cdot 32$ ,, 18° | $-0 \cdot 50$ (red selenium) ($-1 \cdot 3$ melted) |
| Tellurium . . | $-0 \cdot 32$ ,, 18—440° | $-2 \cdot 1$ and $-1 \cdot 6$ ($-0 \cdot 6$ at 18°) |
| Iodine . . . | $-0 \cdot 39$ ,, 18—164° | |
| Bromine. . . | $-0 \cdot 38$ ,, 18° | $-1 \cdot 4$ at 19° |
| Zirconium . . | $-0 \cdot 45$ ,, 18° ($-0 \cdot 3$ at 1150°) | |
| Sulphur . . . | $-0 \cdot 48$ ,, 18—300° | $-0 \cdot 77$ to $-0 \cdot 9$ |
| C (Diamond) . | $-0 \cdot 49$ ,, 18—500° | |
| Chlorine. . . | $-0 \cdot 59$ ,, 16° at atmos. pressure | |
| Boron . . . | $-0 \cdot 8$ ,, 1100° ($-0 \cdot 71$ at 18°) | |
| Phosphorus (white) . | $-0 \cdot 88$ ,, 18° | $-1 \cdot 6$ at melting-point |
| Antimony . . | $-0 \cdot 94$ ,, 18° | $-5 \cdot 6$ to $-3 \cdot 8$ |
| Bismuth. . . | $-1 \cdot 4$ ,, 18° ($-0 \cdot 04$ at 273—405°) | $-16$ at $-182$° ($-14$ at 15°) |
| C (arc carbon). | $-2 \cdot 0$ ,, 18° ($-1 \cdot 5$ at 1150°) | |
| C (graphite) . | | $-8$ at 18° |

In general, the Explanatory Note below Table XXIII applies to this table. In both tables, where no temperatures are given, it may be assumed that no appreciable variation from laboratory temperatures occurred. The unbracketed values, as in Table XXIII, are principally the highest ones given in the Landolt-Börnstein compilation.

*Physical Tables.* While a selection in each case of practically the highest value given in the aforesaid table is made, higher values are possible in some cases under chosen conditions of maximum susceptibility. Practically all the mass values of the feebly magnetic elements are those by Honda. Since the specific gravity of a substance is the weight of a unit volume of the substance compared, the susceptibilities may be converted into mass values by dividing the volume values $k_v$ by the density of the element

$=k_m.$ The suffixes are not always used, as it is understood from the context which magnitude is intended. The volume values in the tables may not be exactly convertible into the mass values given, as the figures relate to distinct experiments often conducted by different experimenters.

Mention should be made of the elements of the rare-earth group, as it has been supposed that some of these in a pure state will be appreciably or even strongly magnetic, judging from the relatively high susceptibilities of some of their oxides of the type $R_2O_3$. The work of B. Urbain and G. Jantsch [*] on these oxides indicates positive values for $k \times 10^3$ as follows :—

$$
\begin{aligned}
\text{Neodymium} &= 33\cdot5 \\
\text{Samarium} &= 6\cdot5 \\
\text{Europium} &= 33\cdot5 \\
\text{Gadolinium} &= 161 \\
\text{Terbium} &= 237 \\
\text{Dysprosium} &= 290
\end{aligned}
$$

Lanthanum is diamagnetic and praseodymium is paramagnetic, whilst holmium has a very high susceptibility, and erbium, thulium, ytterbium and lutecium are successively lower, the values and comparisons being for the oxides. The two last-named elements have the values respectively 18·3 and 3·78 for their oxides according to S. Meyer.[†] This experimenter states that the metallic powders of yttrium, erbium and holmium (mixtures) were not much more magnetic than their corresponding oxides, but that they showed distinct *remanence*, and highly magnetic alloys may be yet possible with some of the elements of this group.

Honda and Soné [‡] in an investigation of erbium (powder), find the value to be 22·4, so that this element would come next to manganese in Table XXIII. Their investigation of the following elements will be of interest : Diamond (dodecahedron) = $-0\cdot452$ ; diamond (octahedron) = $-0\cdot438$ ; sulphur (rhombic) = $-0\cdot476$ ; sulphur (stick) = $-0\cdot444$ ; sulphur (powder) = $-0\cdot416$ ; manganese = $+9\cdot7$ to $+10\cdot1$ ; [§] rubidium (regulus) = $+0\cdot088$ ; selenium (metallic) = $-0\cdot304$ ; osmium (powder) = $+0\cdot074$ ; graphite (Ceylon No. 1) = $-12\cdot2$ to $+0\cdot28$, depending upon its axial positions, etc., in the field. At a temperature of 1200° the susceptibility of the graphite was reduced to one-fifth of its original value ; on cooling below 600° it began perceptibly to decrease, and at 390° zero susceptibility was reached, finally becoming $+6\cdot8 \times 10^{-6}$ at laboratory temperature. The change at 600° corresponds roughly with the magnetic transformation temperature of iron, pointing to the

---

[*] *Comptes Rendus* (1908), 147, p. 1286.
[†] *Akad. Wiss. Wien Sitz. Ber.* (1908), 117. 2A, p. 995.
[‡] *Imp. Univ. Tôkohu Sci. Reports* (1913), ii, No. I, p. 25.
[§] T. Ishiwara (above reports (1916), v, No. I, p. 53) obtains the value $9\cdot66 \times 10^{-6}$ for pure manganese.

presence of iron as an impurity. For diamond, graphite, sulphur, manganese and osmium the susceptibility was found to be independent of the field strength.

The susceptibility of solid oxygen in two forms is of particular interest. A. Perrier and H. K. Onnes * at the Physical Laboratory, Leiden, find that oxygen besides appearing blue-grey also occurs in a transparent form of a vitreous character. The change from the transparent to the blue-grey modification takes place at a temperature of about $-225°$ C. It was noted that throughout the range from freezing-point down to about $-240°$ C. the susceptibility was less than in the liquid state. At about $-240°$ C. the susceptibility suddenly falls to a half-value $79 \times 10^{-6}$, the value before this change takes place being $166 \cdot 1 \times 10^{-6}$. Two changes occur in the susceptibility, viz.—

    (i) A drop to one-third at freezing-point, and
    (ii) A drop to one-half at $-240°$ C.

Oxgyen therefore affords an example of a substance that follows Curie's law at a high temperature, but deviates from it on approaching the absolute zero.

Curie's law states that the mass susceptibility ($k^m$) varies inversely as the absolute temperature — that is to say, $I/H$ is inversely proportional to the absolute temperature. This law is somewhat limited in its application, but it applies to many paramagnetic substances down to extremely low temperatures.

Recent investigations by T. Soné † give the volume susceptibilities $\times 10^6$ for certain gases at $0°$ C. and 760 mm. pressure as follows :—

| | | | |
|---|---|---|---|
| Air | $+0 \cdot 0308$ | Carbon dioxide | $-0 \cdot 00083$ |
| Oxygen | $+0 \cdot 148$ | Pure nitrogen | $-0 \cdot 00033$ |
| Hydrogen | $-0 \cdot 000178$ | Argon | $-0 \cdot 010$ |

In the phenomenon of magnetic susceptibility it may be conducive to clearness to note that the term *ferromagnetic*, as applied to iron, cobalt and nickel (magnetite, $Fe_3O_4$ or $FeO \cdot Fe_2O_3$, is ferromagnetic, and ferro-cobalt, $Fe_2Co$, is also strongly magnetic), serves to distinguish these elements or substances from practically all others which are feebly magnetic or which do not exhibit certain magnetic properties peculiar to iron.

Furthermore, the term *paramagnetic* may be applied to all the elements of Table XXIII, though by virtue of the subdivision just defined the term is frequently used in a restricted sense to designate the feebly magnetic elements of *positive* magnetic susceptibility standing in an opposite relation to those elements which are *diamagnetic*, especially since the latter are without exception feebly diamagnetic.

---

* *Konink. Akad. Wetensch. Amsterdam Proc.* (1914), 16, p. 894.
† *Tôkohu Univ. Sci. Reports* (1919), 8, p. 115 ; *Phil. Mag.* (1920), 39, p. 305.

**A Geometrical Representation of Variable Magnitudes.**—As will be seen from the values given in Table XXIII, a variation from 0·03 to over 31 million is surprising if only some minute structural change in the atom is responsible for this variation. It is probable that this is the case. Manganese steel, it will be remembered, which contains only 12 per cent of manganese and 1 per cent of carbon is practically non-magnetic. It is true that this steel is about fifty times more magnetic than liquid oxgyen (to quote a popular illustration), and magnetised liquid oxygen in comparison with magnetised iron or ordinary steel is very feebly magnetic ; but in commercial use one would say that it is non-magnetic. Thus it may be said that very small internal changes in the atom profoundly modify its magnetisability and even may induce diamagnetism.

Looking now at the problem from a mathematical point of view, one might ask whether there is any geometrical construction which would show such enormous changes with only a small change in certain dimensional parts.

In the case of determining the joint ohmic resistance of parallel conductors there is an equation which has a geometrical construction, viz.—

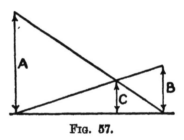

Fig. 57.

where, referring to Fig. 57, the lengths **A** and **B** represent the resistances of corresponding conductors $A_1$ and $B_1$, and **C** is the resulting resistance when they are joined together, as shown by Fig. 58.

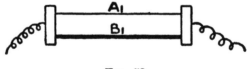

Fig. 58.

This construction, however, does not represent any very considerable variation of the lengths **A**, **B** and **C** for a relatively small alteration in any one of them.

The same general construction, however, lends itself to a variation without alteration in principle, which shows how it is possible for one of the values to become very great with only a slight alteration in the others. This is shown by Fig. 60. In this figure it will be seen that if **C** is increased only slightly **A** will become of infinite length. Thus it will be seen that, provided the

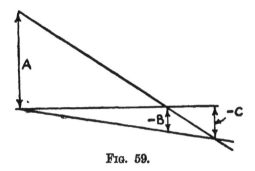

FIG. 59.

values are in proper geometrical relation, there can be produced *one* value of any magnitude.

The equation for Fig. 57 is—

$$\frac{A \times B}{A + B} = C,$$

and for Fig. 59 it is—

$$\frac{-B \times A}{-B + A} = -C.$$

Fig. 60 is in reality Fig. 59 upside down. It, however, serves to show how **A** can become infinite by a small increase of **C**. The

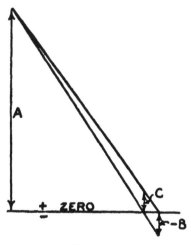

FIG. 60.

letters of this figure are not on the same corresponding lines as those of Fig. 59. It is obvious that Fig. 59 can be drawn so that —**C** will approach infinity.

These examples simply serve to show how it is possible to apply geometrical constructions, or mathematical equations, to give expression to phenomena; and when these are successful new laws or relations are thereby discovered. In the case of magnetism, it is possible that some such construction as illustrated will be found and the elements of the construction identified with known facts. At any rate, these figures serve to show that the assumption of only small inter-atomic or sub-atomic variations producing large magnetic effects, may be based upon mathematical analogy.

**Ferromagnetism and Atomic Structure.**—Sir J. A. Ewing's paper read before the *Royal Society* on 26 May 1921, is of particular interest. It will suffice to give here the abstract of this paper taken from the *Proceedings* of the Society.

" The author refers to his paper published by the Royal Society in 1890,* which modified Weber's theory of the magnetisation of ferro-magnetic substances by showing that the control exerted on each elementary magnet by its neighbours was competent to explain the phenomena of susceptibility and hysteresis. In the present paper this theory is reconsidered in the light of (1) modern views regarding the structure of the atom, and (2) the X-ray analysis of crystal structure by Sir W. H. and Prof. W. L. Bragg. From the results of that analysis it appears that the rotatable Weber magnet is an attribute of the atom, not of the chemical molecule. It is not the atom itself, but probably an electron system within the atom. Metallic iron is now known to be an aggregate of crystals in each of which the space-lattice is the centred cube, with its atoms most closely grouped along the triagonal axes. It is along these axes that the Weber elements will point. Consequently an iron crystal is not magnetically isotropic.

" An important characteristic of the magnetising process in iron is that there is a small quasi-elastic or reversible part preceding the much larger changes which involve hysteresis. This corresponds to a reversible deflection of the Weber magnets through a small angle, generally of an order of 1° or less. The theory of the equilibrium of a row of magnets is considered, and experiments are described in which rows of Robison magnets with ball ends have their equilibrium upset by an extraneous field. The results of these experiments confirm the theory. From the known value of the magnetism of saturation, the moment of a single Weber element is estimated, and the field is calculated which would break up rows of magnets, set in the space-lattice close enough together to bring the reversible deflection within the above limit. The field so calculated is far better than the field that suffices to produce strong magnetisation in iron, which suggests that the ordinary

* See Ewing's book (*loc. cit.*), or *Roy. Soc. Proc.* (1890), 48, p. 342 ; *Phil. Mag.* (1890), 30, p. 206.

laws of force between magnetic elements cease to apply at inter-
atomic distances."

S. J. Barnett read an interesting paper on " The Electron Theory
of Magnetism," before the American Association for the Advance-
ment of Science (Section B), which is given in *Science* (1920), 53,
p. 465. This paper refers to the various contributions to the
subject.

# CHAPTER XXIV

## SIR J. J. THOMSON'S VIEWS OF MASS, ENERGY AND RADIATION

SIR J. J. THOMSON has developed a new method of representing physical processes in terms of moving entities and electric lines of force. An account of Thomson's views first appeared in *Engineering*,\* being part of a report of a series of lectures delivered at the Royal Institution, 1920. These views are presented below. They are based upon the above report and they also correspond closely with a similar one compiled by the writer.†

Thomson's aim is to give a picture of processes of nature which are difficult to visualise, and inasmuch as the ideas are essentially atomic they form a fitting conclusion to the subject of atomic theories and allied subjects discussed throughout these pages.

Now that the atoms are to be regarded as having been built up with some *unit* material, this circumstance becomes of fundamental importance, and the interpretation of physical phenomena in one way would include such a process as a necessary consequence.

In the case of a moving electron its mass is proportional to the square of its charge, and this mass is entirely electric. This electric mass is considered to be diffused throughout the space surrounding the electron, but with a concentration which increases towards its centre, becoming very great close up to it.

The amount of this mass per unit volume an any point P is regarded as being proportional to the square of the electric force at P. Now the electrostatic potential energy at the point P is taken also to be proportional to the square of the force. Moreover, this potential energy is equal to the kinetic energy which the mass would possess if it moved with the velocity of light. This, Prof. Thomson thought, was very suggestive, for it implied that what was called *potential energy* might be the *kinetic energy of the mass*, which, as stated above, was supposed to be distributed throughout the whole electric field.

Generalising, it might be supposed that all mass, whether atomic or electronic, has the same origin and is in reality due to a diffused mass throughout the electric field which surrounds atmospherically all atoms and electrons.

\* 2 April 1920 ; see also *Phil. Mag.* (1920), 39, p. 679 (June).
† *Chem. News* (30 April 1920).

Now the distribution of this atmospheric mass through space is determined by the distribution of the lines of electric force, and all kinds of energy, whether kinetic, potential, radiant, chemical, or thermal, are of one and the same kind, namely, the kinetic energy of this mass which is supposed to be moving with light-velocity.

According to this view, when the energy of a system passes from one kind into another, as from *kinetic* into *potential*, there is no transformation of fundamental energy, but merely the flow of some of this *mass-producing material*, or *fundamental material*, with its intrinsic kinetic energy from one position of space to another under the guidance of the lines of electric force. Thomson is of the opinion that if any kind of reality were to be attached to the term *energy*, the transformation from one kind into another would involve philosophical difficulties ; considering energy as something real, there is a difficulty in forming a physical concept of the transformation process and a violation of the principle of continuity seemed to be involved. To get over this difficulty Thomson introduces the following postulates :—

(1) That mass is made up of exceedingly small particles all of the same kind and very small compared with even an electron.

(2) That these particles move with the velocity of light.

(3) That the force exerted on one of these particles, either by the electric field or by other particles, is at right angles to the direction in which the particle is moving, so that although the path of a particle might be deflected, its energy remains constant.

(4) That each of these units has equal mass and equal energy, and

(5) That their distribution depends upon the number or concentration of lines of force.

It is then assumed that the mass of the particles per unit volume may be evaluated, at the same time retaining the usual electromagnetic equations for the tensions and pressures along and at right angles to the lines of force.

The particles, though moving with the velocity of light, are kept within their small circles of activity by the lines of force which are themselves anchored to the positive particles and negative electrons of the electrified body. These exceedingly small particles, therefore, revolve round the lines of force with the velocity of light.

When, however, the electrified bodies are moving about, the lines of force move with them, and they might be thrown into loops—that is to say, into closed curves without beginning or end ; and when the lines of force are no longer anchored, and can no longer hold the particles in place, they move away with them at full speed, which is the velocity of light. *This hypothesis represents the mechanism of radiation.*

Radiation, therefore, consists of mass-producing material travelling with the velocity of light and carrying with it closed lines of electric force.

It follows, as a consequence, that the emission of radiant energy must be accompanied by a diminution in the mass of the radiating body. This mass-producing material, it will be seen, plays a part analogous to the æther, but there are essential differences. The material is not uniformly distributed throughout the universe, but it is concentrated in the places where the electric forces are greatest—that is to say, round electrons and atoms. Prof. Thomson says : " On this view radiation carries, as it were, its own æther, and when there is no radiation passing there is no æther."

The fundamental material, composed of the particles under consideration, involves functions which were discharged separately by matter and by æther on the old view. The material at one and the same time gives matter its mass and it is the vehicle of radiation ; although the material is the source of mass and energy, it is not the whole of *matter*, i.e. atoms and electrons ; nor is it the whole of *radiation*, as the lines of force are an essential part of both and form the skeleton around which the complete structure is built. The ends and beginnings of these lines of force are the places where the electrons are localised, together with positive particles which build up the atoms and molecules, whilst, in radiation, it is the closed line of electric force which the material carries along with it which produces the polarity essential for the explanation of the properties of light.

An illustration will make this hypothesis clearer. Assume two oppositely-electrified particles A and B, thus—

```
        A                               B
  -  -  •  -   -   -   -   -  •  -  -
        +                               —
```

The mass-producing material is distributed round A and B ; it would tend to flow from the weak parts of the electric field to the stronger. The weak parts being on the sides of the particles nearest to the right and left margins of the page in the accompanying diagram ; and, as a consequence, A would move towards B and B would gain momentum towards A, the two particles thus moving towards each other. Moreover, the material would come from the outlying regions of the field, where it is regarded as potential energy, and in settling round the moving charges it would then appear as kinetic energy.

Thomson believes that along such lines we could get a way of visualising many physical processes and it is an advantage to have some realistic concept of what is happening, and this is more satisfying than mere mathematical formulæ ; indeed, this is recognised by pure mathematicians themselves. " For his own part," Prof. Thomson remarks, " the treatment of physical phenomena as merely mathematical problems took away much of its interest."

In conclusion, it was added that the building up of the universe out of lines of force and small particles rotating round them with the speed of light represent ideas which are consistent with many of the results obtained by the Theory of Relativity.

In this account the writer has followed very closely the excellent report of Prof. Thomson's lecture in *Engineering*.

It will be seen that this hypothesis has a bearing on the *Quantum Theory*, and it will doubtless be discussed by physicists in due time. It is sufficient to record it here.   It should, however, be noted that Prof. Thomson has always been a strong advocate of discrete lines of force, and the new theory as outlined here is apparently a development of Prof. Thomson's existing ideas about radiation as being intimately bound up with lines of force.

# APPENDIX I

## THE ELEMENTS: NAMES, SYMBOLS, ATOMIC WEIGHTS AND ATOMIC NUMBERS: CLASSIFICATION OF THE RADIO-ACTIVE ELEMENTS

### TABLE XXV

| At. No. | Name. | Symbol. | At. Wt. | Isotopes or single values thus far revealed by the positive-ray method. |
|---|---|---|---|---|
| 1 | Hydrogen | H | 1·008 | **1·008** |
| 2 | Helium | He | 4·00 | **4** |
| 3 | Lithium | Li | 6·94 | 6, **7** |
| 4 | Beryllium * | Be | 9·02 | **9** |
| 5 | Boron | B | 10·82 | 10, **11** |
| 6 | Carbon | C | 12·005 | **12** |
| 7 | Nitrogen | N | 14·008 | **14** |
| 8 | Oxygen | O | 16·000 | **16** |
| 9 | Fluorine | F | 19·0 | **19** |
| 10 | Neon | Ne | 20·2 | **20**, 22 |
| 11 | Sodium | Na | 23·00 | **23** |
| 12 | Magnesium | Mg | 24·32 | **24**, 25, 26 |
| 13 | Aluminium | Al | 27·0 | **27** |
| 14 | Silicon | Si | 28·1 | **28**, 29, (30) |
| 15 | Phosphorus | P | 31·04 | **31** |
| 16 | Sulphur | S | 32·06 | **32** |
| 17 | Chlorine | Cl | 35·46 | **35**, 37 |
| 18 | Argon | Ar | 39·9 | 36, **40** |
| 19 | Potassium | K | 39·10 | **39**, 41 |
| 20 | Calcium | Ca | 40·07 | **40**, (44) |
| 21 | Scandium | Sc | 45·1 | |
| 22 | Titanium | Ti | 48·1 | |
| 23 | Vanadium | V | 51·0 | |
| 24 | Chromium | Cr | 52·0 | |
| 25 | Manganese | Mn | 54·93 | |
| 26 | Iron | Fe | 55·84 | (54), **56** |
| 27 | Cobalt | Co | 58·97 | |
| 28 | Nickel | Ni | 58·68 | **58**, 60 |
| 29 | Copper | Cu | 63·57 | |
| 30 | Zinc | Zn | 65·37 | (**64**, 66, 68, 70) |
| 31 | Gallium | Ga | 70·1 | |
| 32 | Germanium | Ge | 72·5 | |
| 33 | Arsenic | As | 74·96 | **75** |
| 34 | Selenium | Se | 79·2 | 74, 76, 77, 78, **80**, 82 |
| 35 | Bromine | Br | 79·92 | **79**, 81 |
| 36 | Krypton | Kr | 82·92 | 78, 80, 82, 83, **84**, 86 |
| 37 | Rubidium | Rb | 85·45 | **85**, 87 |
| 38 | Strontium | Sr | 87·63 | |
| 39 | Yttrium | Yt | 89·03 | |
| 40 | Zirconium | Zr | 90·6 | |
| 41 | Niobium † | Nb | 93·1 | |

* Or Glucinium Gl.      † Or Columbium Cb.

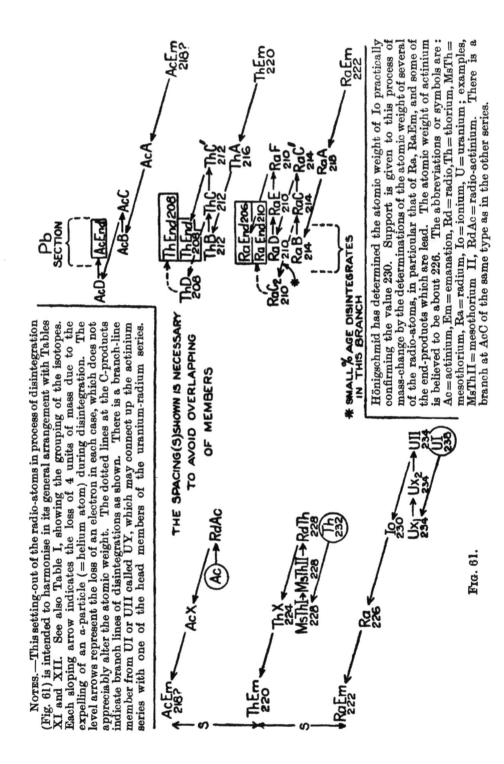

NOTES.—This setting-out of the radio-atoms in process of disintegration (Fig. 61) is intended to harmonise in its general arrangement with Tables XI and XII. See also Table I, showing the grouping of the isotopes. Each sloping arrow indicates the loss of 4 units of mass due to the expelling of an α-particle (=helium atom) during disintegration. The level arrows represent the loss of an electron in each case, which does not appreciably alter the atomic weight. The dotted lines at the C-products indicate branch lines of disintegrations as shown. There is a branch-line member from UI or UII called UY, which may connect up the actinium series with one of the head members of the uranium-radium series.

THE SPACING(S) SHOWN IS NECESSARY TO AVOID OVERLAPPING OF MEMBERS

* SMALL % AGE DISINTEGRATES IN THIS BRANCH

Hönigschmid has determined the atomic weight of Io practically confirming the value 230. Support is given to this process of mass-change by the determinations of the atomic weight of several of the radio-atoms, in particular that of Ra, RaEm, and some of the end-products which are lead. The atomic weight of actinium is believed to be about 226. The abbreviations or symbols are: Ac=actinium, Em=emanation, Rd=radio, Th=thorium, MsTh= mesothorium, Ra=radium, Io=ionium, U=uranium; examples, MsThII=mesothorium II, RdAc=radio-actinium. There is a branch at AcC of the same type as in the other series.

FIG. 61.

TABLE XXV—(*continued*)

| At. No. | Name. | Symbol. | At. Wt. | Isotopes or single values thus far revealed by the positive-ray method (exclusive of emanations). |
|---|---|---|---|---|
| 42 | Molybdenum | Mo | 96·0 | |
| 43 | | | | |
| 44 | Ruthenium | Ru | 101·7 | |
| 45 | Rhodium | Rh | 102·9 | |
| 46 | Palladium | Pd | 106·7 | |
| 47 | Silver | Ag | 107·88 | |
| 48 | Cadmium | Cd | 112·41 | |
| 49 | Indium | In | 114·8 | |
| 50 | Tin | Sn | 118·7 | [116, 117, 118, 119, 120, 121, 122, 124] |
| 51 | Antimony | Sb | 121·8 | 121, 123 |
| 52 | Tellurium | Te | 127·5 | |
| 53 | Iodine | I | 126·92 | 127 |
| 54 | Xenon | Xe | 130·2 | 124, 126, 128, **129**, 130, 131, 132, **134**, 136 |
| 55 | Cæsium | Cs | 132·81 | **133** |
| 56 | Barium | Ba | 137·37 | |
| 57 | Lanthanum | La | 138·9 | |
| 58 | Cerium | Ce | 140·25 | |
| 59 | Praseodymium | Pr | 140·9 | |
| 60 | Neodymium | Nd | 144·3 | |
| 61 | | | | |
| 62 | Samarium | Sa | 150·4 | |
| 63 | Europium | Eu | 152·0 | |
| 64 | Gadolinium | Gd | 157·3 | |
| 65 | Terbium | Tb | 159·2 | |
| 66 | Dysprosium | Dy | 162·5 | |
| 67 | Holmium | Ho | 163·5 | |
| 68 | Erbium | Er | 167·7 | |
| 69 | Thulium | Tm | 169·9 | |
| 70 | Ytterbium * | Yb | 173·5 | |
| 71 | Lutecium | Lu | 175·0 | |
| 72 | Celtium | Ct | | |
| 73 | Tantalum | Ta | 181·5 | |
| 74 | Tungsten | W | 184·0 | |
| 75 | | | | |
| 76 | Osmium | Os | 190·9 | |
| 77 | Iridium | Ir | 193·1 | |
| 78 | Platinum | Pt | 195·2 | |
| 79 | Gold | Au | 197·2 | |
| 80 | Mercury | Hg | 200·6 | (197–200), 202, 204 |
| 81 | Thallium | Tl | 204·0 | |
| 82 | Lead | Pb | 207·19 | |
| 83 | Bismuth | Bi | 209·0 | |
| 84 | Polonium | Po | | |
| 85 | | | | |
| 86 | Emanations | RaEm † | 222 | 220 =ThEm  218 ? =AcEm |
| 87 | | | | |
| 88 | Radium | Ra | 226 | |
| 89 | Actinium | Ac | | |
| 90 | Thorium | Th | 232·2 | |
| 91 | Uranium X2 | UX2 | | |
| 92 | Uranium | U | 238·2 | |

* Or Neoytterbium Ny.                              † Or Niton Nt.

EXPLANATION OF TABLE.—Figures in round ( ) brackets are provisional. Figures in square [ ] brackets are not quite whole numbers. Values in heavy-face type represent, in the case of isotopes, the member in greatest abundance. Heavy-face type is also used for the *single* values.

# APPENDIX II

## MASS AND WEIGHT

THE expressions Weight of one gramme, Conservation of Weight and Atomic Weight are quite accurate statements when it is understood that these are in effect *synonyms* for a constant force value referred to, or taken at, a certain place ; namely, in the case of the ' gramme standard ' a force of 981 dynes involving Paris as the standard place of reference—very much as the boiling-points of substances are referred to a certain atmospheric pressure, or accelerations of gravity are specified in, or reduced to, sea-level values. The actual weight, however, would obviously vary slightly from place to place just as the boiling-point of a given substance would vary if taken at different altitudes ; but fortunately the change in weight is not a troublesome matter, since by the use of a counterpoise balance it is easy to duplicate the quantity of matter represented by the standard mass.

Electromagnetic mass has to be considered in certain atomic phenomena, hence the term *quantity of matter* used above may be interpreted in a specialised or electrical sense. This electrical interpretation of mass arose out of the fact that a moving electric charge (charged body) was found by Thomson (1881) to behave exactly as if it was made more massive by the charge it carried. This subject will be considered later on.

It is of interest to consider the orthodox view of mass which might be defined in the following words, to make use of a definition the writer compiled together with an illustration of the acceleration of mass in a gravitational field.*

*Mass* is a term for the quantity of matter in a body. In order to measure mass, it is assumed that the attraction of the earth on all particles of matter is the same whatever may be their physical state or arrangement, and this attraction is also the same irrespective of their chemical nature. This assumption is fully justified by the fact that bodies of all shapes and of all kinds and sizes fall with equal velocity in the exhausted receiver of an air-pump. This comparative equality of velocity when different bodies are dropped together holds true at all latitudes, though both the centrifugal force due to the rotation of the earth and the gravitational attraction vary according to the latitude of the place. If a one-pound mass

* *Chem. News* (1915), 111, p. 301.

at a given place fell slower than a two-pound mass at the *same* place, one would, upon dividing the latter into two one-pound parts, have to account for a miraculous disappearance of force ; hence the equality in velocity arises from a given force acting with a given amount of matter. Therefore it is possible to measure the mass of a body by its weight, using for this purpose a counterpoise balance, and this mass can be defined as a quantity proportional to the weight, in the sense that if at a given place on the earth's surface one body is twice as heavy as another the mass of the first is twice that of the second. Suppose, however, that the body be weighed by a spring-balance at a certain place removed from the Equator, and weighed again by the same instrument at another place on or near to the Equator, it will be found that the body is lighter at the latter place. It is found, too, that the acceleration due to the force urging the body towards the earth is also less at the second place than at the first, in the same proportion. This illustrates the fact that when the mass remains the same the weight varies as the acceleration of gravity.* Hence the weight varies as the product of the mass and this acceleration, and consequently, when suitable units are chosen, the mass of a body is equal to its weight divided by the acceleration due to gravity. Acceleration is the rate of increase (or change) of velocity per unit of time (see below).

Going over familiar ground, it is of course known that the *kilogramme* is defined as the mass of a piece of platinum deposited for safe keeping and reference in the Archives of Paris, and that the standard unit in the *Centimetre-Gramme-Second System* is a *thousandth* part of this mass = *one gramme*.

At Paris the acceleration of gravity ($g$), as it is termed, is such that the gramme mass (or any other mass) will fall there under the influence of the earth's attraction, when dropped from a height, with a velocity of 981·0 cms. per second at the end of the first second, at the end of the second second $2 \times 981$, at the end of the third second $3 \times 981$, and so on.

Suppose that a cinematograph picture be taken of a falling watch against a stationary measuring rule or scale 256 ft. long. It will be found that at the 16-ft. mark on the rule the watch registers one second in advance of its time when liberated. The watch has taken one second to cover 16 ft. During consecutive seconds it will have traversed 16, 48, 80, and 112 ft. respectively (neglecting fractional values), the total distance traversed being the sum of these. If the watch had attached to it a speed-indicating mechanism, actuated, say, by a pinion engaging with a rack alongside the

* The term *gravity*, as thus used, is not a pure term, but involves the diluting effect of centrifugal force. This does not matter, since all forces can be added to or subtracted from one another as if they were of the same kind. Modern ideas have led, in fact, to a monistic doctrine whereby they merely become circumstantial variations of the same general phenomenon. See, for example, A. Einstein, *Relativity : The Special and General Theory*, translated by R. W. Lawson.

rule, then it would be found that just at the end of each successive second it had indicated a velocity of 32, 64, 96, and 128 ft. per second respectively.

This will be understood from the following analysis :—

The velocity at the *end* of 16 ft. must be greater than 16 ft. per second, taken as an instantaneous velocity at this point ; it is 32, as we shall see in the next paragraph.

The acceleration being *uniform*, it can be said that 16 ft. per second represents a mean value ; hence the maximum velocity, occurring at the end of the first second, would be $16 \times 2 = 32$ *ft. per second.*

Similarly, at the end of the second second the total distance traversed would be $16 + 48 = 64$ ft. This also being a mean value for two seconds, the maximum velocity then becomes $64 \times 2 = 128$ ft. per two seconds. Since it is only a question of the velocity per second at the end of the second second, this would be $128/2 = 64$ *ft. per second.*

To continue : at the end of the third second the total distance covered from the start would be $16 + 48 + 80 = 144$ ft. The maximum velocity $= 144 \times 2 = 288$ ft. per three seconds. The time taken being three seconds, the velocity per second at the end of the third second would be $288/3 = 96$ *ft. per second* . . . and so on.

Now, during each successive second the gain per second has been 32 ft., since

$$\left.\begin{array}{l} 32 - 0 = 32 \\ 64 - 32 = 32 \\ 96 - 64 = 32 \end{array}\right\} = g$$

so that it may be said that the *acceleration per second per second is* 32 *ft.*, and that it is *uniform*. The exact figure is practically 32·2 ft. per second per second at London, or 981 cms. per second per second at Paris.*

This means that the force acting downwards on the gramme mass is 981·0 dynes, because force = mass × acceleration, the *dyne* being the unit of force which causes unit acceleration in unit mass ; or, from another point of view, a unit force of 1 dyne is that force which, acting on unit mass for unit time, causes it to move with unit velocity. It is of course understood that in this case the unit

---

* It is of possible interest to note that the distances 16, 48, 80, 112, etc., stand to each other as the numbers 1, 3, 5, 7, etc., while the velocities at the ends of the consecutive seconds stand to each other as the numbers 1, 2, 3, 4, etc., and the distances ($D$) from the beginning (each time taken) are the squares of the numbers 4, 8, 12, 16, etc.

$D = \frac{1}{2}gt^2$.   $t$ = time in seconds.

$t = \sqrt{D} \times \frac{1}{4}$.   Final velocity in feet per sec. = $\sqrt{D} \times 8$.

8 is the square root of $2g$.

Mechanical friction due to the speed-indicator, as well as to air friction, is of course taken as a negligible quantity in discussing main principles.

of distance or length is the centimetre, the unit of mass is the gramme, and the unit of time is the second.

Since *force = accceleration × mass*, and as the mass remains constant—in this case, unity—981 × 1=force of gravity, as it is termed. The acceleration is also constant; therefore the force must have acted with the same pull at the outset—that is, before the mass started on its downward movement; hence the force is the weight of the stationary mass of 1 grm. expressed in dynes.

The term *quantity of matter* is frequently avoided in precise scientific literature as it is said to lack definition; but substituting the word 'body' for it only conveys to the practical mind the idea of a quantity of matter in ordinary mechanics. It is, in short, the most expressive and the best definition that can perhaps be formulated, but like many others it is defective since it is based upon a somewhat conventional idea of matter. Now the monistic doctrine (to use a term employed in this connection by Soddy) that there is, only one kind of mass which is fundamentally electric has gained acceptance of late amongst many physicists.

The electrodynamics of $\beta$-particles has been carefully studied (see work of Kaufmann and that of Bucherer) and it is found that there is actually a variation of mass with velocity as there should be if the mass is of electric origin. The formula used, which agrees with experiments, is that of Lorentz, viz.—

$$m_v = m_o(1-\beta^2)^{-\frac{1}{2}}$$

in which

$m_v$=the electromagnetic mass due to the high-speed movement of the $\beta$-particles.

$m_o$=the mass of the $\beta$-particle when not moving quickly : at a speed less than $\frac{1}{10}$th that of light.*

$\beta$=the ratio of the velocity of the $\beta$-particle to that of light.

The following quotation taken from Einstein's article in *Nature* (1921), 106, p. 782, is of interest in this connection : " It was found that inertia is not a fundamental property of matter, nor, indeed, an irreducible magnitude, but *a property of energy* [the italics are the present writer's]. If an amount of energy E be given to a body, the inertial mass of the body increases by an amount, $E/c^2$, where $c$ is the velocity of light *in vacuo*. On the other hand, a body of mass $m$ is to be regarded as a store of energy of magnitude $mc^2$.

" Furthermore, it was soon found impossible to link up the science of gravitation with the special theory of relativity in a natural manner. In this connection I was struck by the fact that the force of gravitation possesses a fundamental property,

* Strictly speaking, the velocity should be zero, or the mass considered at right angles to the line of motion of the $\beta$-particle.

which distinguishes it from electro-magnetic forces. All bodies fall in a gravitational field with the same acceleration, or—what is only another formulation of the same fact—the gravitational and inertial masses of a body are numerically equal to each other. This numerical equality suggests identity of character. Can gravitation and inertia be identical ? This question leads directly to the General Theory of Relativity. Is it not possible for me to regard the earth as free from rotation, if I conceive of the centrifugal force, which acts on all bodies at rest relatively to the earth, as being a ' real ' field of gravitation, or part of such a field ? If this idea can be carried out, then we shall have proved in very truth the identity of gravitation and inertia. For the same property which is regarded as *inertia* from the point of view of a system not taking part in the rotation can be interpreted as *gravitation* when considered with respect to a system that shares the rotation. According to Newton, this interpretation is impossible, because by Newton's law the centrifugal field cannot be regarded as being produced by matter, and because in Newton's theory there is no place for a ' real ' field of the ' Koriolis-field ' type. But perhaps Newton's law of field could be replaced by another that fits in with the field which holds with respect to a ' rotating ' system of co-ordinates ? My conviction of the identity of inertial and gravitational mass aroused within me the feeling of absolute confidence in the correctness of this interpretation. . . . The following are some of the important questions which are awaiting solution at the present time. Are electrical and gravitational fields really so different in character that there is no formal unit to which they can be reduced ? Do gravitational fields play a part in the constitution of matter, and is the continuum within the atomic nucleus to be regarded as appreciably non-Euclidean ? A final question has reference to the cosmological problem. Is inertia to be traced to mutual action with distant masses ? And connected with the latter : Is the spatial extent of the universe finite ? "

## REFERENCE

J. A. Crowther, *Ions, Electrons and Ionising Radiations*, pp. 198–202.

# APPENDIX III

## LEAD ISOTOPES

ISOTOPIC leads have been obtained in sufficient quantity to enable their atomic weights and other properties to be measured accurately. The agreements or otherwise in certain properties may be stated as follows :—

1. Atomic weights differ appreciably.
2. Spectra practically identical (see Chapter II).
3. Densities slightly different.
4. Specific volumes slightly different (see p. 129).
5. Atomic volumes and diameters alike.
    > Soddy gives * the results of experiments on the atomic volumes of lead isotopes, pointing out that it seems safe to conclude that these volumes cannot differ from one another by so much as three parts in ten thousand; moreover, that the atomic diameters in such cases cannot differ by so much as one part in ten thousand.
6. Water dissolves amounts by weight of radium D nitrate and ordinary lead nitrate in proportion to their respective molecular weights.
7. Atomic number the same.

Generally, chemically and physically the isotopes are alike, and where a difference appears it is very small, as in the case of 3 and 4. In the case of 6 the proportionality is determined by the molecular weights. This has been found to hold true with uranio-lead nitrate with respect to ordinary lead nitrate.†

8. The melting-points of lead isotopes agree within 0·06 to 0·01 degree C. of one another, which is within the range of experimental error.‡

---

* *Nature* (1921), 107, p. 41.
† See T. W. Richards, *Nature* (1919), 103, pp. 74 and 93.
‡ See T. W. Richards and N. F. Hall, *Am. Chem. Soc. Journ.* (1920), 42, p. 1550.

9. The X-ray absorption wave-lengths are as follows : *—

| Isotope | Atomic weight | $\lambda \times 10^8$ cm. for critical-abs. wave-lengths | | |
|---|---|---|---|---|
| Ordinary lead . . . . | 207·19 | 0·9485 | 0·8128 | 0·7806 |
| Lead of radio-origin . | 206·08 | 0·9489 | 0·8129 | 0·7810 |

It thus appears that the critical-absorption wave-lengths of lead isotopes are the same within the range of experimental error. These results show that the relative change for the three critical wave-lengths are the same for each isotope.

10. The refractive indices of isotopic lead salts have been determined.† Lead isotopes of atomic weight 207·19 and 206·41 combined with nitrogen to form crystals of lead nitrate give refractive indices :—

1·7815 and 1·7814 respectively,

each value being the mean of a number of concordant determinations. Thus there is no difference that could be attributed to anything other than the slight variations in these experiments.

* Duane and Shimizu, *Nat. Acad. Sci. Proc.* (1919), 5, p. 198.
† T. W. Richards and W. C. Schumb, *Am. Chem. Soc. Journ.* (1918), 40, p. 1403.

# APPENDIX IV

## BALMER SERIES OF SPECTRUM LINES

BALMER in 1885 showed that the lines of the ordinary hydrogen spectrum could be calculated by means of a simple formula. Expressing the lines in wave-lengths the following equation is of historical interest :—

Wave-length in cm. $=\lambda=\{M \times m^2/(m^2-4)\} \times 10^{-8}$.*
$M$ being a constant $=3646{\cdot}14$.
$m$ being successive integers of a series, 3, 4, 5, etc.

Other formulæ have been proposed as shown under the head of Rydberg Constant in Appendix V.

According to Bohr's theory (see Chapter X), it is supposed that the attenuated hydrogen of nebulæ exists in a sufficiently low state of pressure as to permit of much larger electronic orbits (larger atoms in consequence) than would be the case in the laboratory, since Bohr's equation involved in its interpretation atoms of different size. The sizes are supposed to vary as the second power of the integer numbers of a series thus : $1^2$, $2^2$, $3^2$, $4^2$ . . . etc. Consequently in the nebulæ and stars additional lines would be present, and this is the case.

Professors T. R. Merton and J. W. Nicholson † have shown that with various gaseous media through which an electric discharge was passed various intensities of the Balmer series of lines could be obtained. In Table XXVI these results are shown. The first column is the particular wave-length line of hydrogen ($\lambda$). An arbitrary intensity of 100 is assigned to the H$\alpha$ line in each case. Some of the figures are here 'rounded off' to obviate too many decimal fractions.

On the whole these results seem somewhat difficult to reconcile with Bohr's theory, especially since it was found that with a mercury pressure of 41 mm. (D) it was possible to observe 12 lines of the Balmer series.‡

* Omitting this 'multiplier' would give the value in Ångström units, as one Å unit $=0{\cdot}00000001$ cm.
† Roy. Soc. Proc. (1919), 96, p. 112.
‡ See R. T. Birge, Phys. Review (1921), 17, p. 589. Birge remarks on the possibility of fields of force produced by impurities (such as oxygen, which is more easily ionised by collision than some other elements) present in gaseous mixtures causing a better production of the Balmer series.

The authors remark : " The transfer of energy to members of lower-term number in water-vapour at 5 to 10 mm. pressure, and in helium at 41 mm. pressure, would appear to accord with Bohr's theory which demands the existence, for the production of higher members, of a considerable number of greatly enlarged atoms " ; but this appears to be outweighed by the circumstance that members of the series having a high-term member ($m=14$) are visible at higher pressures. "It seems probable that assumptions of a far-reaching character will be needed to explain the present results in the light of Bohr's theory."

TABLE XXVI

| $\lambda$ | Water vapour 7 mm. pressure of mercury (A) | Pure hydrogen less than 1 mm. pressure of mercury (B) | Mixture of hydrogen and helium less than 1 mm. of Hg (C) | Traces of hydrogen in helium 41 mm. pressure of Hg (D) |
|---|---|---|---|---|
| H$\alpha$ | 100 | 100 | 100 | 100 |
| H$\beta$ | 23 | 42 | 60 | 32 |
| H$\gamma$ | 5·9 | 17·0 | 14·0 | 7·0 |
| H$\delta$ | 1·0 | 3·9 | 2·4 | 1·0 |
| H$\epsilon$ | just visible | 1·2 | — | 0·18 |

R. W. Wood * has succeeded in extending the Balmer series of hydrogen lines as far as the 20th line. Hitherto the 12th line was the laboratory limit, as the continuous light background completely masked any faint lines present, and besides secondary lines † appeared which obscured those of the proper series. By means of a vacuum-tube with an elongated portion which is used in an end-on position, these new lines were photographed. The 20th line has an intensity $\frac{1}{800,000}$th of that of the H$\alpha$ line. This further experimental work supports Bohr's theory in some respects, yet questions arise which are difficult to reconcile with it ; or rather, some further elucidation seems necessary. On the whole the theory seems to be fairly well established (see Birge, *loc. cit.*).

* *Roy. Soc. Proc.* (1920), 97, p. 455.
† These secondary lines are probably due to the hydrogen molecule.

## THE RYDBERG CONSTANT

THE constant, as used in the type of equation given by Bohr (Chapter X), in which it has the value $3\cdot290 \times 10^{15}$ is not, of course, the same as the true Rydberg constant which is a different value, being used in such formulæ as here given.

W. E. Curtis * has recently compiled the following formulæ in which—

$$B, k, \beta\dagger \text{ and } \mu \text{ are constants,}$$

and $m$ is a number of a series of successive integers 2, 3, 4, 5, 6, etc.

The first three formulæ are of the Rydberg type.

The fourth is due to Bohr, with a small correction for the change of mass of the electron with velocity, and

The fifth and sixth are due to H. S. Allen.

$$\text{Wave No.} = \frac{N}{(2+\beta)^2} - \frac{N}{(m+\mu)^2} \qquad , N = 109678\cdot28 \quad . \quad . \quad \text{(i)}$$

$$\mu = +0\cdot0_5210 \text{ and } \beta = -0\cdot0_5383.$$

$$\text{Wave No.} = \frac{N}{4} - \frac{N}{(m+\mu)^2} \qquad , N = 109678\cdot73 \quad . \quad . \quad \text{(ii)}$$

$$\mu = +0\cdot0_574.$$

$$\text{Wave No.} = \frac{N}{(2+\beta)^2} - \frac{N}{m^2} \qquad , N = 109678\cdot10 \quad . \quad . \quad \text{(iii)}$$

$$\beta = -0\cdot0_553.$$

$$\text{Wave No.} = N\left(\frac{1}{4} - \frac{1}{m^2}\right)\left\{1 + k\left(\frac{1}{4} + \frac{1}{m^2}\right)\right\}, \quad N = 109677\cdot58 \quad . \quad . \quad \text{(iv)}$$

$$k = +0\cdot0_4387.$$

---

* *Roy. Soc. Proc.* (1919), **96**, p. 147.
† In Curtis' paper, this symbol ($\beta$) was indicated by $\mu$

$$\text{Wave No.} = \frac{N}{4} - \frac{N}{\left(m + \dfrac{B}{m^2}\right)^2} \qquad , N = 109678\cdot76 \quad . \quad . \quad \text{(v)}$$

$$B = +0\cdot0_4629.$$

$$\text{Wave No.} = \frac{N}{\left(2 + \dfrac{B}{4}\right)^2} - \frac{N}{\left(m + \dfrac{B}{m^2}\right)^2} \qquad , N = 109677\cdot79 \quad . \quad . \quad \text{(vi)}$$

$$B = -0\cdot0_43144.$$

The relations of some of the quantities involved may be expressed thus—

$\lambda =$ wave-length in centimetres.

$c =$ velocity of light in centimetres per second, $= 3 \times 10^{10}$ ; or taking a more exact value $2\cdot9986 \times 10^{10}$, these being the velocity in a vacuum.

$T =$ periodic time in seconds.

$\nu =$ frequency per second.

Wave No. $=$ number of waves per centimetre.

To convert the wave number into the frequency value ($\nu$) as obtained by the equation (see Chapter X)—

$$3\cdot290 \times 10^{15}\left(\frac{1}{q^2} - \frac{1}{p^2}\right) = \nu$$

the wave-number may first be divided into $10^8$ (since Wave No. $= 10^8/\lambda$) to convert it into wave-length ($\lambda$).

Now as $\quad \lambda = cT,$

Therefore $\dfrac{\lambda}{c} = T$ ;

Consequently, $\dfrac{1}{T} = \nu.$

Note.—The wave-numbers given by equations (i) to (vi) have not to be multiplied by $10^8$ to get the number of waves per centimetre; and similarly, $10^8/\lambda$ above should be taken as $1/\lambda$ when not working in Angstrom units; but the usual wave-length figures have to be multiplied by $10^{-8}$ to give the wave-length in cm. measure.

# APPENDIX VI

## THE RUTHERFORD-SODDY DISINTEGRATION LAW

RADIO-ACTIVE phenomenon is essentially the disintegration of the (radio-)atom by steps, so that for every $\alpha$-particle (=helium atom with a double positive charge) emitted the parent atom loses 4 units of mass and this resulting atom thus becomes one of less atomic weight. Uranium has an atomic weight of 238·2. Between this element and lead there are 12 direct-line changes, corresponding to 12 elements, but all of very transitory existence except 4 which are long-lived. The helium atoms given out by eight changes, counting that of uranium itself, represent a mass of $8 \times 4 = 32$; so that lead from this line of descent has an atomic weight of $238·2 - 32 = 206·2$. This reduction in mass whereby the element is changed to that of another type has been fully proved by experiments which are in perfect agreement with each other, and there is an accumulation of evidence that this law is rigorously true. The emission of $\beta$- and $\gamma$-rays enters into the problem when $\alpha$-rays are not emitted, but these rays do not appreciably alter the atomic weight though a chemical change takes place when the $\beta$-rays are emitted, as has been already explained in Chapter IV.

# APPENDIX VII

## THE MOLECULE

THE term *molecule* connotes an atom, or more generally a compound atom, as Dalton termed it, which has a *free existence*. For example, the *monatomic molecules* are : Na, K, Zn, Cd, Hg, He, Ne, Ar, Kr, Xe, Ems., I, Br, etc. The *diatomic molecules* are : $H_2$, $N_2$, $O_2$, $F_2$, $Cl_2$, $Br_2$, $I_2$, $S_2$, $Se_2$, $Te_2$, $As_2$, $Bi_2$, etc. The *triatomic molecules* are : $H_3$, $O_3$, etc. The *tetratomic molecules* are : $P_4$, $As_4$, etc. The *pentatomic molecules* do not appear to be recorded. The *hexatomic molecules* are : $S_6$, etc. The *septatomic molecules* do not appear to be recorded. The *octatomic molecules* are : $S_8$, etc. The " *etcs.*" (last four) here used indicate that there may be a few more molecules of the respective types given.

There is a tendency for the chemically-active molecules having atomicities of 1, 3, 5 and 7 to be particularly unstable. This instability seems traceable to the cubical symmetry of the electronic arrangement, as will be seen from the combinations of atoms as indicated by the octet theory. The metals are only monatomic at high temperatures and when they are in the liquid and solid states they are ionised, according to the views given by Langmuir (see Chapter XIV). Ozone is particularly unstable. $H_3$ has as yet only a vacuum-tube existence.* The other higher members of this *odd* series do not appear to exist. The inert gases are monatomic but they belong to the *even* series : 0, 2, 4, 6 and 8.

Now molecular formulæ, such as NaCl, do not represent a definite molecular state in the solid, as the Na and Cl atoms exist as ions and arrange themselves so as to share in all directions their affinities. This condition is indicated by the spacing of the atoms revealed by X-ray analysis (see Chapters VI and VII). However, in the process of combination such a formula as NaCl gives the correct combining proportions, but concurrently with or immediately after combination the ionic state occurs in which no molecular identity appears, judging from the crystalline structure of such salts.

Some molecules have an existence as free units without marked affinity so long as their atoms remain in combination, as instanced by $H_2$, $O_2$, $N_2$, etc.

* See pages 22 and 48.

Apart from the electrical field round the atom which gives rise to chemical combination, the structural stability of the molecular unit may be such as to admit of the easy separation of the atoms, and then they would be free to form new combinations with other atoms or molecules present.

X-ray analysis has revealed the structure of matter so that in some cases there are no distinct molecules in a solid substance, the atoms of the substance being alternatively placed so as not to represent any definite molecule : unless the crystal as a whole be regarded as a vast molecule. In other cases, the X-ray analysis has revealed the close packing of units consisting of more than one atom which if free to move about could constitute molecules in the sense of isolated and free units. The structure of ice is an example. Citing the most recent investigation, that by D. M. Dennison,* it was found from the positions of 12 clearly-defined lines that the lattice of an ice crystal corresponds to hexagonal molecules in a closely-packed arrangement comprising two sets of interpenetrating triangular prisms with side measurements of 4·52Å units and a height of 7·32Å units. With close-packed spheres the axial ratio should be 1·633 ; whereas from the data obtained in this investigation the ratio is 1·62, thus indicating a flattening of 0·8 per cent which was in the direction of the hexagonal axis. This lattice is practically the same as that found for magnesium,† but the relative intensities of the two X-ray patterns indicate that the magnesium atoms must differ considerably in shape from those of the ice molecules. From the density of the ice and the lattice dimensions it appears that the ice molecule is $(H_2O)_2$ or $H_4O_2$, *this being the molecular formula for ice.*

Passing now to water vapour, this substance may be represented by the formula $H_2O$, being monomolecular. It is of interest to note that slight apparent deviations from this formula may arise from the neglect of applying corrections for the gas laws which are not perfect (see Chapter I).‡

Though the inert gases crystallise, at least in the case of argon, according to the experiments of W. Wahl,§ there is as yet no evidence that they associate to form simple molecular complexes in the solid state similar to water when frozen.

Molecular phenomena should be studied from the point of view of the octet theory as the mechanism of molecular and atomic states is elucidated by this theory. At the same time the work of Rutherford in partly breaking up certain atoms affords information as to the atomic structures, and this has to be taken into account in developing any molecular theory.

---

* *Phys. Review* (1921), 17, p. 20 ; or *Chem. News* (1921), 122, p. 54.
† See Wyckoff, *Am. Journ. of Science* (1920), 50, p. 317.
‡ See J. Kendall, *Am. Chem. Soc. Journ.* (1920), 42, p. 2477 ; or *Chem. News* (1921), 122, p. 30.
§ *Roy. Soc. Proc.* (1913), 87, p. 371.

There is an important set of considerations touched upon above which arise out of the work on crystal lattices, viz. :—

1. *Lattice structures* which lead to *molecular* lattices as distinct from *atomic* lattices.
2. *Atomic lattices* which give no definite molecular characteristic to the compound.

While 1 and 2 have been taken into account in developing the octet theory, it remains to be seen whether modifications in existing views may not be necessary as this field of investigation is developed further.

R. W. G. Wyckoff,* in an article on the "Determination of the Structure of Crystals," remarks : " In spite of the obvious characteristics that pervade all these discussions there seems to be considerable evidence, from a consideration of the crystal structures of such compounds as have been more or less carefully studied, for the existence of two distinct types of solid compound—(1) *polar compounds* [ions], wherein the bonding between the atoms, or at least between certain groups of atoms, is polar ; and (2) *valence compounds*, the atoms of which are bound to other atoms by holding electrons in common. A consideration of organic compounds, none of which have thus far been successfully studied using X-rays, forces us to a third kind of compound ; (3) the *molecule-forming compounds*, built up of groups of atoms (the chemical molecules) held together presumably by relatively weak stray fields of force. It will be noticed that in solids of the first two types no molecules in the chemical sense of the word exist : each crystal is a single chemical individual."

* *Franklin Inst. Journ.* (1921), 191, p. 199.

# APPENDIX VIII

## NEGATIVE INDICES

FOR the benefit of those readers not accustomed to mathematical expressions the negative index will be explained.

When values are very small it is not convenient to write them out. Take for example : $0.000000000000075$.*

By means of the $^{-n}$, or, negative index, this figure may be expressed thus—this form of expression affords another advantage, as given below—

$$75 \times 10^{-15}$$

since $10^{-15} = 0.000000000000001$, and this decimal fraction multiplied by 75 equals the first-given figure above.

Now to understand what $10^{-15}$ means, the simplest plan is to 'work it out' by starting a series from the general expression—

$$N^{-n} = \frac{1}{N^n},$$

so that assuming $N = 10$

$$10^{-1} = 0.1 \; ; \; \text{or} \; \frac{1}{10^1} = \frac{1}{10} = 0.1$$

$$10^{-2} = 0.01 \; ; \; \text{or} \; \frac{1}{10^2} = \frac{1}{100} = 0.01$$

$$10^{-3} = 0.001 \; ; \qquad \text{etc. etc.}$$
$$10^{-4} = 0.0001$$

and so on in a similar manner. The number of noughts to the right of the decimal point is one less than the index figure.

To appreciate more fully the negative indices it is better to start 'positively' with, say, $10^3 = 1000$ and work back past zero index to, say, $10^{-3} = 0.001$, thus—

$$10^3 = 1000$$
$$10^2 = 100$$
$$10^1 = 10$$
$$10^0 = 1$$
$$10^{-1} = 0.1$$
$$10^{-2} = 0.01$$
$$10^{-3} = 0.001$$

* This figure could be represented thus $0.0_{13}75$—a practice sometimes resorted to.

14

It is of interest to note that indices can be added provided the number N is the same, and it is desirable to use multipliers of 10 to the $n$th power whether positive or negative as regards the index.

For example—

$$10^6 \times 10^{-2} = 10^4 = 10000,$$

or

$$10^0 \times 10^{-1} = 10^{-1} = 0\cdot 1$$

The rationale of the $^0$ power or index will be seen by virtue of the above series—see also below.

Other series of the above general type are equally instructive. For example—

| | | |
|---|---|---|
| $2^0 = 1$ | $\sqrt{16} = 4$ | $1^0 = 1$ |
| $2^1 = 2$ | $\sqrt{4} = 2$ | $2^0 = 1$ |
| $2^2 = 4$ | $\sqrt{1} = 1$ | $3^0 = 1$ |
| $2^3 = 8$ | $\sqrt{\cdot 25} = 0\cdot 5$ | $4^0 = 1$ |
| | $\sqrt{\cdot 0625} = 0\cdot 25$, etc. | Any figure to the $^0$ power equals *one*. |

Note the regularity of the differences between successive values, which affords a criterion for the validity of the series.

A negative sign may accompany a fractional power, thus

$$10^{-\frac{a}{b}}$$

which means by analysis that the $b$-root of 10 has to be taken, thus—

$$\sqrt[b]{10}$$

and this has to be raised to the $a$-power, thus—

$$(\sqrt[b]{10})^a$$

and then the whole divided into *unity*, thus—

$$\frac{1}{(\sqrt[b]{10})^a}$$

If, therefore, the values are $a=1$ and $b=2$, then

$$10^{-\frac{1}{2}} = \frac{1}{\sqrt{10}}$$

The $a$-value being 1 it may be, obviously, neglected, as raising $\sqrt{10}$ to the *first* power does not alter it.

NOTE.—10 may be raised to the $a$-power *first*, so that

$$10^{-\frac{1}{2}} = \frac{1}{\sqrt{10^1}}$$

or, by way of further examples—

$$10^{-\frac{3}{4}} = \frac{1}{\sqrt[4]{10^3}} = \frac{1}{(\sqrt[4]{10})^3} = 10^{-0\cdot 75}.$$

# NAME INDEX

# SUBJECT INDEX*

* NOTES.—Cross-references are given in the chapters. Name Index will be
found useful in tracing the work of leading scientists.